DEAD MILE

'A brilliant cop, a highly entertaining story and a smart new take on the
"locked room" mystery make for a winning combination. I loved it!'
T.M. LOGAN, AUTHOR OF *THE HOLIDAY*

'Spellbindingly original and utterly compelling. Strap yourselves in . . .
Dead Mile is one hell of a ride!'
JACK JORDAN, AUTHOR OF *DO NO HARM*

'An inventive, gripping take on a locked room mystery'
HARRIET TYCE, AUTHOR OF *BLOOD ORANGE*

'The action is brisk, the characters are vividly drawn, and the plot snakes
around the throat. A tense and ingenious "locked-traffic" thriller'
VASEEM KHAN, AUTHOR OF THE MALABAR HOUSE SERIES

'An ingenious, high-octane page-turner with twists and turns
worthy of a Formula 1 course. Utterly brilliant'
KATE SIMANTS, AUTHOR OF *FREEZE*

'A gripping, grid-locked masterclass in how to how to ratchet
up the tension! I devoured it in one sitting!'
ROBERT RUTHERFORD, AUTHOR OF *SEVEN DAYS*

'A pacy page turner with a twisty plot and a sense of tension
that doesn't let up until the final page. Loved it'
NIKKI SMITH, AUTHOR OF *LOOK WHAT YOU MADE ME DO*

Jo Furniss is originally from the UK, but spent much of her adult life overseas, living in Cameroon, Switzerland and Singapore. She's now back home, beside the seaside in England. Her novels include the survival thriller, *All the Little Children*, which was an Amazon Chart bestseller in the UK and US. Jo has contributed short stories to the *Afraid of the Light* anthology series, which was nominated for the CWA Dagger Awards in 2021/22, and raised thousands of pounds for British charities. A former BBC broadcast journalist, Jo also writes for the award-winning *Short History Of* podcast from Noiser Productions. Visit her at JoFurniss.com or follow her on Instagram and Twitter @jofurnissauthor.

DEAD MILE

JO FURNISS

ZAFFRE

First published in the UK in 2024 by
ZAFFRE
An imprint of Zaffre Publishing Group
A Bonnier Books UK Company
4th Floor, Victoria House, Bloomsbury Square, London WC1B 4DA
Owned by Bonnier Books
Sveavägen 56, Stockholm, Sweden

A CIP catalogue record for this book is
available from the British Library.

ISBN: 978-1-80418-344-1

Also available as an ebook and an audiobook

1 3 5 7 9 10 8 6 4 2

Typeset by IDSUK (Data Connection) Ltd
Printed and bound in Great Britain by Clays Ltd, Elcograf S.p.A.

Zaffre is an imprint of Zaffre Publishing Group
A Bonnier Books UK Company
www.bonnierbooks.co.uk

For my criminal friends

Pincers up!

THE FIRST HOUR

Chapter One

5.00 p.m.

Friday afternoon, and the traffic was bloody murder. Belinda 'Billy' Kidd should have known better than to take the motorway at rush hour. Plus she wasn't feeling tip-top after a long-haul flight from Australia. As a police sergeant – soon to be *former* police sergeant – she was well aware that 'a can of Red Bull and I'll be fine' could lead to 'but I only rested my eyes for a moment' and carnage.

But what choice did she have? There was no welcoming party at Arrivals. No public transport to her village. So here she was in a hire car. Aircon set to Arctic. Black coffee in the centre console. White knuckles on the steering wheel.

It should only take an hour to drive home. One little hour. An easy drive she'd done a million times. A road she'd patrolled back in the day, though it had been upgraded since then to a three-lane 'express' route that clogged up every Friday into a joyless conga, plodding towards the weekend. Billy hovered in the middle lane even though it wasn't moving any faster than the others. She'd like to put the radio on but couldn't bring herself to glance away from the road long enough to work the unfamiliar stereo. The only sound was the drumroll of wheels.

Wheels skidding, tyres bursting—
Stop it, Belinda.

On cue, red crosses lit up the overhead signs. Brake lights haemorrhaged down the carriageway. She rolled to a standstill. The engine put itself to sleep.

Silence revealed her ragged breath.

Over thirty years behind the wheel. Advanced police driver qualifications. One horrible incident. Now this; heart thumping like a windscreen wiper on full pelt.

She'd had a proper panic attack the first time she drove on a freeway in Australia. Her sister told her it's the menopause. Lots of women lose their confidence in mid-life. *Get on the HRT*, Mel said, *you'll be back to your old self, Sergeant Billy Kidd, scourge of scamps and scoundrels and scumbags.* But she didn't drive again on Aussie asphalt. And even now, back on British tarmac, she was at a standstill on a Friday afternoon and thoroughly relieved about it.

How can you be a police officer if you can't drive?

Oh, give it a rest, Belinda.

If someone spoke to her the way she spoke to herself, she'd tell them to stick it up their exhaust pipe.

In the outside lane to her right, a silver fox in a BMW hammered his steering wheel. In the inside lane to her left, two women stared ahead with identical profiles. Mother and grown-up daughter. Thirty years difference made them look like a before-and-after advert. She followed their gaze up the carriageway. Stationary as far as the eye could see.

Billy broke the other-worldly silence by clicking on the radio.

'Scores dead in a series of coordinated attacks—' The presenter's voice sounded both grave and gleeful. 'Following a car bomb explosion outside the train station at 4.30 this afternoon, gunmen opened fire inside the concourse. Transport hubs across the city are being evacuated. Police have barricaded roads leading to and from the city centre, which is effectively locked down—'

The roar of a motorbike engine outside on the carriageway threatened to drown out the news. She cranked up the volume on the radio.

'—and reports are coming in now of a second car bomb. There was an explosion in the Deadwall Tunnel at 5 p.m. precisely. This is looking to be the most deadly incident on British soil since the London bombings on the seventh of July 2005—'

No wonder the motorway was at a standstill. The Deadwall Tunnel was the main route into the city from the east. Billy wasn't sure exactly where they were – there was a high barrier on both sides of the motorway so she couldn't look for landmarks – but the tunnel could only be a mile ahead. She dropped her face into her hands. Outside, a door clunked and Billy's head snapped straight back up.

In her mirrors, she watched drivers emerge from their vehicles to peel sweat-darkened shirts from their backs, make phone calls, turn forlorn circles like lost tourists.

'What a nutter!' The voice of a young woman standing in the outside lane.

Sound carried in the strange stillness left by the absence of road noise. The girl, no more than twenty years old, wore a shimmery, fish-skin frock and no shoes. Barefoot as though the earth were her red carpet. Her glinting dress and sandy-coloured tan made her look like a mermaid. In response to her shout, a lad with the bohemian brunette curls of Lord Byron jumped out of his VW camper van. He sauntered on puppy feet that he hadn't grown into yet towards Mermaid's Nissan Micra.

Billy looked away from these beautiful young things.

In the inside lane, a sturdy old girl got out of her car to glare accusingly up and down the hard shoulder. Her lips moved through conspicuously foul-mouthed words. Billy's sister would describe Foul Mouth as 'a unit'.

The traffic wasn't going to move any time soon, not with the tunnel blown up and the city locked down. So Billy got out too, dragging her jet lag like a spent parachute.

The air reeked of hot rubber.

Tyres skidding—

Don't, Belinda.

She'd brought the weather with her. The car's gauge said twenty-eight degrees, which was, what, eighty-two in old money? A heatwave in May. The opportunity for British people to complain about relatively moderate heat came earlier every year. Maybe the whole bloody country was menopausal?

Billy glanced around the flushed faces of the other drivers. Behind her in the middle lane, an expensive-looking woman – caramel highlights, statement necklace, ginormous SUV – fanned herself with a parking ticket. She caught Billy's eye.

'Can't believe this heat,' Parking Ticket said. 'Today of all days.'

Billy wondered why she didn't take off her long-sleeved cardigan then.

'It'll end in a storm,' Billy said, but the sun winking in a pastel blue sky suggested otherwise.

'This stretch is always busy of a Friday, but it's not usually this bad,' Parking Ticket mused.

'There was a car bomb in the Deadwall Tunnel.'

'Christ almighty!'

'We're lodged in the throat of a bottleneck. We'll have to wait 'til it coughs us out.' Billy grabbed her phone and went to Maps. Right enough, this stretch of motorway was cut off by the Deadwall Tunnel way ahead. Behind them was a long stretch of road that was unbroken by an exit for about the same distance. And the traffic going in the other direction had stopped now too, so the whole system must be snarled up.

'Could we drive out?' Parking Ticket waved at the hard shoulder. 'If it's a personal emergency?'

'You wouldn't get far; the tunnel is blocked. And driving on the hard shoulder is illegal.' She pointed at the surveillance cameras. 'You'll get another ticket. Sorry, I'm a police officer.'

Parking Ticket threw the parking ticket into her car and walked to the shade at the back of her enormous vehicle. Conversation over.

Bit rude. Who needs a car that size anyway?

An Audi the size of an Aldi.

Interrupting that thought, an abrupt scream.

Parking Ticket was backing away from the vehicle that had stopped alongside hers in the inside lane. A black sedan. Distinctively anonymous, an unrecognisable brand. It reminded Billy of the *X-Files*.

Why was Parking Ticket pointing at it?

She waggled her finger between Billy and the sedan. 'You said you're police?'

'I'm almost retired.' Billy stayed beside her car.

'He isn't moving,' said Parking Ticket. A pink flush slid up her neck like fingers hidden inside her shirt.

It was true that the driver of the black sedan was awfully still.

His vehicle was skew-whiff, its nose turned towards the hard shoulder. Sunlight smeared the windscreen. Billy couldn't see the driver's face as she gave in and started to approach. He seemed to be slumped. She tapped his window. No response.

'Hello?' she called out. 'Are you alright, sir?'

No response.

Hardly likely he was taking a nap – the traffic only stopped a few minutes ago.

Billy cupped her eyes and pressed against the window. Yeah, he wasn't well. Slack lids and lips. Waxy-white eyes. Young guy, hardly old enough for a heart attack, but that was Billy's best guess.

'Can someone call an ambulance?' she yelled over her shoulder.

She had to get this man out of the vehicle, administer CPR.

Parking Ticket wailed. 'Is he . . . ?'

'Dead,' said Foul Mouth from where she stood at the nose of the sedan, her arms folded atop a shelf of her own bosom. 'I'm a nurse and that one's dead.'

Billy lifted the driver's door handle. It released, but she hesitated to fling the door open. Deep down she already knew. Even through glass, she could tell that the life had gone out of this man.

Chapter Two

5.05 p.m.

Billy opened the door of the black sedan and the driver's arm slithered down the front of her legs before hanging still. She stared a moment at his limp hand to make sure it had stopped, then angled her upper body inside the car to place two fingers on the man's throat.

No pulse.

She reached around to feel the other side of his neck.

No pulse.

A wet patch encircled the groin of his jeans, but there was no smell. Urine still fresh. She talked to him as she pulled her phone from her back pocket, switched on the torch and lifted one of his eyelids.

No reaction.

With thumb and index finger, she found the trapezius muscle at the top of his right shoulder, took a pinch of flesh through his shirt and twisted. Hard. A pain stimulus would get a reaction if a reaction could be had.

No reaction.

Billy backed out of the car and stood up.

'Did you say you're a nurse?' she called to Foul Mouth. The woman stood at the bonnet of the black sedan staring steadily through the windscreen.

'I was, yeah.'

'I'm Sergeant Kidd. Belinda Kidd, everyone calls me Billy. Help me get him out?'

'What for?' Foul Mouth said.

'He's not responding. We'll do CPR,' Billy's throat closed on the end of the word. The last time she tried resuscitation, it hadn't ended well for the recipient or her career.

'How long's he been gone?' Foul Mouth regarded the dead man as though she suspected him of some complicated ruse. Billy looked around for someone else to help. CPR was a long shot without a defibrillator or any kind of equipment, but *bloody hell* they could at least try. Parking Ticket had buggered off too. Okay, she'd have to do it by herself, then, but she needed the man out of the car first.

She grabbed him by his collar. Fabric squelched between her fingers. She whipped her hand away from the back of his neck. Kept on feeling that squelch as she inspected a starburst of blood on her palm.

Billy backed out of the driver's side and opened the rear door with her clean hand. The man's neck was framed between the seat and the headrest. A bright stain on his collar formed the shape of a red neckerchief, the kind worn by a cowboy or a shaggy dog. The truth was less jaunty. Blood seeped from a wound to the right of his spine, which had a black metal object protruding about a centimetre from it.

'Told you,' said Foul Mouth, leaning inside the front to feel his wrist. Bending down in the heat made her wheeze. The nurse muttered curses instead of last rites. She emerged to announce, 'He's gone, love.'

In fairness, the thing sticking out of his neck did suggest this wasn't a heart attack.

Billy closed the rear door so that Foul Mouth couldn't eyeball the man's injury. The thing embedded in his nape wasn't visible from the front, there was only the smear of blood to indicate that he was wounded at all. As her mind slid from natural causes to suspicious circumstances, she decided to keep the details to herself. She quickly wiped the blood off her hand onto the underside of her T-shirt.

'Keep the blood between us?' she muttered to Foul Mouth, lowering her voice so it didn't carry across the muted carriageway. 'Don't want people to panic.'

'Whatever you say, boss.'

'We could still try CPR, though,' said Billy.

The man was so newly expired that his last breath must be floating about the interior. She tried not to breathe it in as she checked him for other visible injuries, but there were none. He was maybe thirty, handsome, in good nick. Bit of a waste, really.

Billy had once attended an accident where a woman drove a motorbike through a fence that took her head clean off. Her body was in perfect condition. Apart from the obvious.

Focus, Belinda.

The point was, decapitation was an easy call; dead or not dead – definitely dead. This man was not definitely dead. Yes, he had something in the back of his neck, but had it killed him? There could still be a chance of preserving life. But even if they did CPR, it was pointless unless backup was coming.

'Did someone call that ambulance?' Billy shouted over her shoulder.

She heard a cry of acknowledgement.

Time was ticking away. She turned around to ask the kid who looked like Lord Byron to help her lift the bloke out of the car – with that thing in his neck, she couldn't just tip him onto the ground, they'd have to do it more carefully – but Byron had a phone pressed to his ear. He was the one calling 999. He caught her eye and shook his curls emphatically.

What did that mean?

He lowered the phone, redialled and called again.

Oh, the terrorist attacks.

Of course.

That took the wind out of her sails.

The emergency services would be overwhelmed. Byron couldn't even get through to an operator. It more or less signalled the end of the road for the dead driver. First, he had a thing in the back of his neck. Second, he wasn't breathing. Third, if she could bring him back from the brink, he'd need an ambulance within half an hour max, and if all the services had been diverted to these attacks in the city they wouldn't get here for ages. Four, it was bloody hot and she wasn't sure she could keep pumping his chest for half an hour in this heat. CPR was really physical work . . .

Chalk up one more victim to the terrorists.

She hung in the doorway for a moment.

'You should call it,' said Foul Mouth.

I'm not a bloody heart surgeon.

Billy hated to stand down when she'd rather step up, but what more could she do? If this menopause-driving-phobia-thing forced her to retire (out of necessity and sheer bloody embarrassment) then she would have to get used to sitting on the sidelines.

Totally bloody useless.

Stop.

The counsellor said thoughts can't hurt you.

But what if they can?

Thoughts make people do things. A panic attack is thoughts. A suicide starts with thoughts. Murder even.

She glanced at the dead man.

Fine, she'd keep a tight rein on her thoughts. She could deal with the dead driver, drive home as soon as the traffic cleared – no freaking out – get some sleep, write a letter to accept early retirement, then retrain to do something useful such as groom poodles. She might even get her own poodle. She could do that if she were retired. These were all encouraging thoughts.

But her eyes slid back to the dead man. He was wearing a wedding ring, so shiny it looked new. Who was he? How on earth did he come to die here? How long was it going to take Major Crimes to arrive on the scene? They'd be too busy with these terrorist attacks, hardly enough officers on a normal day, never mind during a crisis.

Her heart pinched and she pressed her fingertips deep into her left boob. When had she lost the professional composure that ate up adrenaline like fuel? Could this really be hormones? Or was it the heat – the air felt charged and unpredictable.

The man's fingers dangled above the tarmac, the colour of uncooked pork sausages. She was tempted to flop them into the car and slam the door, but shouldn't touch anything. When did he die? It was too soon to draw conclusions from the lividity in his fingers – blood takes time to succumb to gravity and drain through the body and pool in the lowest parts. Like these sausage-fingers.

How long since the traffic stopped? That would be a better indication.

The car bomb in the Deadwall Tunnel had gone off on the hour at five o'clock. It had only taken a minute or so for the Friday traffic to back up and gridlock. She checked her watch.

'He's been dead less than ten minutes.'

'How can you be sure it wasn't longer?' asked Foul Mouth, arms still firmly crossed.

Billy allowed a twitch of irritation to show in her eyebrows. 'He must have been compos mentis enough to brake, otherwise he would have slammed into the car in front. So he must have been alive when we stopped.'

A thought barrelled into Billy's brain like a truck on a skidpan. Someone did this to him.

She opened the rear door again, careful not to touch anything. Knelt down on the road to peer up at his neck, framed in the headrest. The thing that protruded from one of the puncture wounds was metal. Black. Thin and round as a kebab skewer. She didn't know what it was, but it was lodged into the nape of the man's neck. Angled up into the brain as though they'd been hiding on the back seat.

Someone had murdered this man. And they did it right here on the motorway.

Chapter Three

5.09 p.m.

Billy looked around the bystanders. The backs of her knees were slippery with sweat as she stood up from inspecting the dead man's neck. Everyone was standing in the outside lane, respectfully – or squeamishly – facing in the other direction. She moved around the SUV to join them. Its owner, Parking Ticket, was being brusquely consoled by Foul Mouth. Mermaid and Byron hovered. Most people had chosen to remain in their air-conditioned cars so hadn't clocked the situation yet. But voices drifted, they'd know soon enough. The drivers of the vehicles in front and behind the black sedan had yet to emerge, which Billy found strange because they must be aware of the drama.

Perhaps a lack of interest was better than rubberneckers. All she had to do was protect the crime scene. Don't get involved. Wait for the professionals. Go home and retire. Despite herself, Billy slipped into police autopilot and got their full names.

'Wish I had some water,' said Mermaid or Daisy Finch. 'It's too hot.'

Byron – Olly Sims – said he had water in the camper van and loped off.

'Where are we anyway?' Daisy pinched the screen on her phone. She held up her device to show a map. Billy seemed to have been appointed 'leader' whether she liked it or not. The satellite map showed a solitary thick grey line of motorway cutting through a huge sandy-brown blob of nothing much.

'Blythe Flats, the old gas works,' Billy said with a nod that indicated the world beyond the motorway. 'Biggest construction site in the country. Or it will be if they ever get permission to build. They put in the Deadwall Tunnel to feed the motorway under the development, then the land was found to be contaminated, and it all got put on hold. It was in the news.'

Olly arrived back and handed the young woman a small flask, with an apology to everyone else that he didn't have more bottles. 'Why is there a fence on both sides? It's really claustrophobic.' His accent made it sound like *rarely* claustrophobic. With his foppish hair and accent, he still reminded Billy of Lord Byron. He just needed a flouncy shirt and a rapier.

A thin metal weapon.

Like the black skewer in the dead man's neck.

Billy tried to put it out of her mind. Not her problem.

'I guess they built the fence to protect the motorway from the construction site,' she said, 'to stop the dust and debris and what have you blowing across the road.'

'It's like a prison wall,' muttered Foul Mouth, better known as Pat Mackey.

On each side of the carriageway, there were three lanes of stationary traffic, plus a strip of hard shoulder in a pinkish tarmac the colour of an internal organ, then a grass verge with a few straggly trees that didn't disguise the fact that they were barricaded in by the high barrier. The fence must be four storeys high. It ran as far as Billy could see. With the traffic at a standstill, there was no way forward, no way back.

'So, basically,' said Daisy, 'there're no services anywhere? I'd better make this bottle last.' She cracked the lid and jumped back as it spilt. The splashes dried before their eyes.

In part, this high fence was why they were so hot. With no cross wind, it was a sun trap. Billy fancied she could see a low fug settling in, tinged by car fumes, yellow and toxic. Her head felt floaty, her ears popped, her breath tasted of tin.

People were going to start feeling this heat. No one trusts a forecast of hot weather in England in May so they'd all come out trussed up in jeans and jumpers. Except Daisy, obviously, in her mermaid dress. The heat would make people either listless or fractious. Billy never underestimated the influence of the weather; crime falls when rain drives folk inside, crime rises when sunshine encourages them outside. Storms mean car crashes. Heatwaves mean drunken clashes. Football means battered wives. Didn't take a criminologist to explain cause and effect to Billy. But how a traffic jam had killed a man – that baffled her.

She peeled away from the group and went to regard the dead driver from the vantage point of the hard shoulder. He seemed to be looking at her. Lids lowered. Shy, almost. She wondered if he'd cried out. Surrounded by people, and nobody heard him. It was like a bad dream.

Retirement or no retirement, she was a police officer and had to act. Her colleagues all over the city were dealing with much worse. Imagine the scenes in the Deadwall Tunnel where the car bomb had gone off. At the very least, Billy could stop trespassers spoiling a crime scene.

First, she went to Pat Mackey, who was rummaging in a carpet bag in the boot of her Volkswagen Golf.

'It's important you don't talk to anyone about the condition of the dead driver,' Billy reminded the nurse. 'That car is a crime scene and anything you saw could be part of a future investigation.'

'It's alright, I watch TV.' Pat spoke without looking up. 'You don't want the cause of death to be made public, so that when some little scrote says "I didn't strangle her", you can say "how did you know she was strangled?", and then you got him.'

'True, but it's more than that. Even the smallest detail. His injuries. His position. His clothing. Anything. As people start to realise what's happened they're going to want to ask you about it, it's human nature to be nosy, but I need you to say nothing. Nothing at all.'

Pat stood up and held a Jiffy bag stuffed with medications to her chest while she gave the black sedan a hard stare. If the woman was going to argue the point, Billy would have to threaten her with obstructing an investigation. But she shifted her gaze to Billy and leant in.

'One of the great skills in life,' said Pat, 'is knowing when to talk and when to shut the eff up. They can ask me what they want, I've dealt with tougher crowds than this lot.'

Billy didn't doubt it somehow.

'So you won't say anything to anyone about what you saw in that car?'

Pat raised three fingers in a salute. 'Girl Guide's honour.'

'Is that a yes?'

'Fuck's sake, yes, no, I won't tell anyone about the blood.'

Billy nodded her thanks and turned to go.

'But you know they can just walk over there and see the dead guy for themselves,' Pat said to her back.

She wasn't wrong. Billy would have to deal with that. She crossed the rows of stranded vehicles towards the drivers still standing in the outside lane. They were chit-chatting about the events in the city centre, avoiding all mention of the elephant in the room. Their conversation faltered as a police officer approached, bringing with her a distinct whiff of elephant.

'Did anyone see anything strange when the traffic first stopped?' Billy asked.

Charlotte McVie – who'd had a rough day, what with getting a parking ticket and finding a dead body – gave a little mewl. 'Only that man, not moving.'

'Anyone walking or running away from the scene?'

'Surely we would have noticed someone on the carriageway?' said Charlotte.

'Yah, you'd notice as you ran the bugger over,' added Olly.

'Not if they were on the hard shoulder, though?' Billy asked.

'They'd be even more conspicuous,' said Charlotte. 'No cover.'

This was true.

'Who was the first to get out of their car?' Billy asked.

'Me,' said Daisy. 'I was sweating cobs. No air conditioning.'

'Did you see anyone else on the road? Ahead or behind us?'

'No, it was just me. And I had a good look round, too, to see what everyone else was doing. It felt weird being out of the car, wrong somehow, standing on the motorway. Like I don't belong. And I remember thinking maybe I'm the first person ever to walk on this bit of tarmac in its entire life, which is weird, isn't it?' Without any pockets in her slinky dress, her hands wound around each other. 'And I thought it might be illegal or something. But then he got out too' – she indicated Olly – 'and the overhead signs said road closed for an incident, so I thought it must be alright.'

Daisy couldn't be more than twenty. Student, probably. Young and uncertain and wearing a nightclub dress during the day. What an age. Old enough to make all the good mistakes and young enough to go back home to your mum if they turn out to be bad mistakes.

Billy didn't let that thought linger too long, motherhood was a rabbit hole she preferred to scamper past. Her own experience of parenting had been short and sweet, and came to an abrupt end a long time ago. Just over twenty years, in fact – she would never forget the date. Even if time was the much-promised healer, the scar was always close to the surface and vulnerable to getting scratched.

'So long as everyone doesn't abandon their vehicles, you're alright,' she said to reassure the girl.

Pat arrived and handed Charlotte a packet of menthol tissues.

'Why are you asking, anyway?' Daisy chewed a hangnail. 'And how did he die?'

'The post-mortem will determine that,' said Billy.

'A post-mortem like in *Silent Witness*?' asked Daisy, hand landing on her heart. 'Oh my days, he's been killed.'

Well, he hardly died in his sleep . . .

'I mean, he was murdered?' Daisy corrected herself with the flicker of a grimace at appearing daft or naive.

Charlotte gave another mewl and turned away.

Pat scrunched her face into a frown but Billy eyeballed her and the nurse did as she'd promised and stayed quiet.

That metal thing didn't get in the bloke's neck by itself. If no one had seen a suspect on the run, then the killer couldn't have left. But Billy was only one person, one officer. Never mind securing a crime scene, how could she protect all these people from a killer?

Chapter Four

5.14 p.m.

Experienced police officers were supposed to present to the world a regulation 'impassive face' that gave nothing away. But keeping a secret had never been Billy's forte. People seemed to sniff intrigue on her as if she was made of cake.

Now, on the motorway, although she hadn't said aloud the words *murder*, *killing*, or *suspicious death*, a rumour that the dead man had been murdered, killed or suspiciously made-dead radiated around the bystanders as though carried on the heat rising from the sun-soaked tarmac. More drivers and passengers emerged, drawn by the drama.

The mother and daughter in the minivan in front of Dead Driver had got out to stand on the hard shoulder. Facially, they were alike, but outside the car a difference was apparent – the girl was as skinny as an arrow. They must have twigged about the dead bloke behind them as the daughter was staring over her shoulder at the sedan. One hand twirled the thin hair on the top of her head. That gesture, and their little-and-large sizes, made Billy think of Laurel and Hardy. The mother tugged the girl's hand to make her look away, and evidently decided they'd be better off sitting in the car where he was out of sight. The daughter took one last glance as she obediently got in her passenger seat.

Someone called out from the other side. The traffic had come to a halt on the southbound carriageway too. The whole system must have snarled up as surrounding roads got gridlocked by vehicles backing up from the lockdown in the city. She told the bloke there was nothing to worry about, nothing to see here, please wait in the car out of the sun. But she heard his under-the-breath instruction to do something to herself. Didn't care about that, she'd heard it all before, but she needed everyone to keep their distance, keep their calm.

The rumour meant the temperature on the already boiling motorway rose another couple of notches. If she couldn't keep control – which was hard enough for an officer in uniform, never mind one in the jeans and T-shirt she'd put on yesterday in Australia – then her crime scene would be compromised. The least she could do for the dead bloke was give him a chance of being investigated. She needed to preserve any evidence.

Billy checked that everyone was at a safe distance and opened his rear door. She couldn't touch anything – no gloves – but took a good look at the black metal skewer in the back of the man's neck. She collected photos with her phone; the body in situ, the cabin of the vehicle, the road around and under the car, even the grass verge beyond and the surrounding traffic, documenting the position of vehicles.

On her screen, she zoomed in on a shot of the man's neck to see the weapon more clearly. It was weird. The object was thin and black with an end that curved ninety degrees to a flattened top like the head of a nail. She had never seen anything like it.

Carefully placing his dangling arm onto his lap, she closed all the doors on Dead Driver. She felt bad, calling him that, but she couldn't find his name without going through his pockets. John Doe, Joe Bloggs, Dead Driver – it didn't really matter, did it? The important thing was finding out what happened to him.

God only knows how the crime had been committed, but it had, and this was the mess the SOCOs would have to make sense of when they arrived. She wouldn't make it worse. She would try, for once, not to make it worse.

She looked around and saw Pat waving a packet of wet wipes.

Oh, God.

I've already made it worse.

She'd let the nurse lean inside the car and manhandle him. DNA all over Dead Driver. 'Foul Mouth' had been furious when the traffic stopped, in a right state, effing and jeffing. What if Pat did it? They'd have no forensics. And she'd seen the blood, she knew there was something up with his neck.

Wake up, woman.

Engage brain.

Billy accepted three wipes from Pat, wondering if she'd been smart enough to insert herself into the crime scene for this very reason. And to gain a police officer's trust. Billy cleaned her hands and under her fingernails.

'This is a crime scene,' she called out to everyone within earshot. 'Please don't approach this vehicle.'

Nods all round. Wide eyes too.

'Has anyone got disposable gloves or plastic bags or anything?'

What she'd give for a well-stocked panda car.

Nitrile gloves. Evidence bags. Police tape.

Baton, pepper spray, taser.

Uniform would help.

Radio.

But she didn't have any of these things so she'd have to improvise.

And be careful.

'I might have some bags,' said Charlotte, and she dived into her SUV.

Billy got her phone out to try again to call it in.

While she waited for the 999 call to connect to an operator, she noted how the stationary vehicles on her northbound side of the carriageway formed a useful grid. In the inside 'slow' lane, from back to front: Dead Driver, little-and-large mother and daughter whose names she still needed to take, and Pat Mackey the foul-mouthed nurse. In the middle lane, the high-maintenance Charlotte McVie in the SUV, Billy's own hire car, and a Renault with baby seats whose driver hadn't got out. In the outside lane, posh Olly Sims's camper van, Daisy Finch's hot-pink Micra, Silver Fox in the Beamer. Silver Fox was approaching now, looking fussy.

'Should we think about moving—?' Silver Fox started.

'She's police,' Pat snapped, rounding on him. 'You're not touching anything without her say-so.'

Silver Fox held up his hands and retreated.

Billy barely had time to establish that his name was Nigel Heathcote.

'Bloody coming over to mansplain it,' muttered Pat.

Billy was still hanging on the 999 call. The operator answered at last.

'Emergency. Which service?'

Billy asked for police, was redirected and walked away onto the grass verge to describe the situation without being overheard by all and sundry. She mentioned that she was a sergeant from a neighbouring police force and needed to report a road traffic collision.

'What's the location of the RTC?' asked the operator. The background hubbub prompted a sparkle in Billy's veins, the sharp adrenaline of a control room in crisis.

'It's not a crash, it's a . . . Well, I'm not sure what it is. The driver is dead. We're all stuck in a traffic jam.'

'Is there a dangerous or excessive build-up of traffic?'

'Yes, we're stuck in a traffic jam.'

'The RTC caused the traffic jam?'

'No, the road was closed because of the car bomb in the Dead-wall Tunnel. Then we realised a driver was dead.'

'Have you attempted CPR?'

'He's got a metal skewer in the back of his neck.'

'He's been stabbed?'

'I don't see how. He was alone inside his car.'

There was a pause. 'It is illegal to make hoax calls. We always prosecute persons who abuse the communications system. It is also illegal to pretend to be a police officer.'

'It's not a hoax. And I am a real police officer.' Billy listed her collar number. 'I understand there are a number of ongoing terrorist incidents and you're very busy, so you probably can't spare a crew right now, but I need to report the death and get help as soon as you can send it.' After being warned that there would be no officers or ambulances for some time, the operator ended the call sounding as though she didn't quite believe her.

Billy looked from the dead man's frozen face to the animated expressions of those all around. All eyes on her. Sometimes people seemed more suspicious of the police than the bloody criminals.

Only Nigel Heathcote stared in the direction the traffic should be travelling. He leant against his passenger door, arms folded, massaging his own biceps. He didn't look like a Nigel. Trendy haircut. Ray-Bans in May. Fancied himself a bit, did the silver fox.

'Found these.' Charlotte approached holding out a packet of scented bags for disposing of dirty nappies. In her other hand, she had a yellow tin of Rescue Remedy herbal sweets. She had the lean frame and knotty face of someone who did hardcore exercise, marathons, maybe triathlons. An excuse to get away from home? The woman retreated to her vehicle and climbed inside. Certainly preferred her own company.

But she must have been friendly enough to give herbal sweets to Daisy and Olly. The lad screwed up his face and pulled the pastille from his mouth, then the girl snatched it and popped it into her own. *Blimey.* How quickly can two people reach saliva-sharing familiarity? Less than half an hour in their case.

Unless, Billy thought, they're not the strangers they would appear to be?

Daisy looked up, sensing Billy's attention.

'Excuse me?' she asked. 'Why are you looking at the back of his head?'

'Making enquiries,' said Billy in her blandest voice.

'Are they sending more crews?' Daisy fiddled with a line of sequins that ran down the front of her skimpy dress, brushing them so they laid flat, making them less conspicuous. But they caught the sun as she moved, like fish giving themselves away under the surface.

The answer was no, there was no one coming to provide support, not with the terrorist attack in progress. But Billy didn't want to announce that to the whole carriageway. Didn't want the killer to know that the only thing standing between him and freedom was one almost-retired policewoman.

'They'll send scenes of crime officers as soon as they're available.'

'Haven't you contaminated the forensics?' Daisy persisted.

'My first job is always to preserve life,' Billy said, 'if I can.' She turned away, signalling that the tutorial was over.

Daisy was about the age that Billy's daughter should have been now. But this was no time to allow a soft spot to develop.

Focus, Billy.

Step away from the rabbit hole.

This girl was someone else's daughter. And she was a bit too interested in the details of the investigation. She had to go on Billy's radar for that reason, and that reason alone.

Billy licked her lips and tasted fuel. Up and down the length of the motorway, wafts of voices and music and even food might suggest this was a picnic in the park. Except it wasn't. A man had been murdered. It was a crime scene.

She hadn't always understood the police officer's compulsion to run into danger when most people would run away. Was it nature or nurture? Bravery or bloody-mindedness? She'd given her life to policing, and in recent years it hadn't given much in return.

But now she felt that familiar tug on the invisible chain of duty.

Maybe it was self-preservation – there was a killer here somewhere – or a desire to prove herself. Maybe it was another weird symptom of the bloody menopause. One way or another, her racing heart had known before her reluctant head that she was always going to run into this particular danger.

Charlotte aka Parking Ticket

Audi Q7 (2023 model, Hybrid SUV)

I catch myself making driving noises through pursed lips. *Brrr-rum-brum.* The twins do the same, when they're allowed in the car.

What if they're crying? What if they're scared? What if all this backfires?

They'll be the ones to suffer.

My hands peel off the steering wheel with a slick sound.

The boys are safe.

Calm your thoughts.

Have to find them first!

His voice in my head. Thinks he's funny, but really he needs to diminish what he doesn't understand or have the patience for and it threatens him that I'm trying to improve myself, at least that's what Jemima says and Jemima has enough certificates on her wall to charge £120 for a forty-five minute consultation, so she should know because she is an absolute lifesaver, quite literally in some cases, even if he says she's not a real doctor just a head doctor who peddles self-pity.

Have a Rescue Remedy.

And of course the lid is stuck. Every. Bloody. Time.

That lad with the hair pushed it on too tight.

'*It is typical of the patriarchy to put a stress-inducing lid on a well-being product for women. Is it simply ineptitude or a rare form of gaslighting?*'

Composing letters of complaint is a way to get through the day.

If I dig my nails under the lid and release the air seal, air lock, vacuum, whatever you call it – yes, the little gasping sound of it coming free is a relief after a tiny skirmish with the patriarchy in which I have come out victorious. Hands are fluttering. Get a grip, literally, get a grip. Get one in your mouth, you foolish woman.

The taste is soothing.

Calm your thoughts.

Must take the tin out of the car after. Mustn't leave anything incriminating. Wipe the interior down too. Unless the car is going in the reservoir. Still haven't decided on that detail. Maybe that plan is like the letters of complaint – stress relief, wish fulfilment, revenge fantasy, whatever, better in concept than reality. I never send any letters, do I? In any case, reservoir or no reservoir, there's no going back. Things should get better now, they have to. Hope is eternal. Is that a song? Hope lies eternal. Hope lives eternal. It's a poem. Or the Bible. Why don't I know these things? He would know, he always bloody knows. Or says he does—

Don't start with that again.

Do something else. Switch on the radio.

Car bombs.

Shootings.

Gridlock.

Turn it off.

Off!

Too late—

I catch myself making the noise again. *Brrr-rum-brum.* And the imp is free. It's here. Trapped inside with me. Panic is a wicked little imp that capers around the cabin snatching the breaths right out of my mouth. Its sharp claws scamper up my shoulder scratching—

No!

Jemima says thinking of your anxiety as an imp puts it out of your control, like it's a wild animal, when it's only your thoughts.

Rotten stupid human woman thoughts.

Have to find them first!

Control yourself, for fuck's sake, you sound just like him.

Breathe.

It's not an imp.

Breathe.

The imp settles.

And look outside. The sky. What colour? Bright blue. Try harder, be specific. Lulworth Blue. That's it, Farrow & Ball call it Lulworth Blue. And the leaves? Faded. More grey than green. Mizzle, that one's called. The road is in Railings, obviously. The policewoman is wearing a white top that's been washed so many times it's like Blackened or Dimpse. She'd look better in something more striking, like Arsenic.

Look at her, the policewoman, where is she going? Poking about in the trees. Where'd she get a tan at this time of year? She's got that look they have, coppers. Move along, nothing to see. Well, what if there is something to see and no one's looking, not even the police? Fuck her. Fuck. Her. Let her nose about in the trees, what does she think she's going to find over there, the fountain of youth?

Oh—

Hope *springs* eternal. That's the phrase.

The traffic will clear soon. We'll be on our way – there is still time.

Chapter Five

5.21 p.m.

Billy wanted to establish the lie of the land beyond the black sedan's final resting place in the inside lane. She crossed the pitted pink tarmac of the hard shoulder onto a wide grass verge where the scraggly trees had snared and shredded plastic bags that her dad used to call witches' knickers. Soggy litter underfoot, a yoghurt carton, a ripped Durex sachet; *how is that even possible?*

Beyond the copse rose the wall of plastic panels. Its pattern was faded to the colour of spilt bodily fluids. The barrier towered over her. There was a matching one on the other side of the motorway. It made her think of a gulag. It made her think this was a dismal place to die. She laid her hand on one of the emaciated silver birch saplings. There is only so much CO_2 these poor trees can suck down. Her mind skipped to those experiments they used to do where beagles were forced to smoke cigarettes.

Focus, Belinda.

The point was, it's no life for a tree, standing beside an English motorway. Pushing through monoxide-grey branches, she reached the plastic wall. The fence was too high, too steep, too slippery to climb. No doors or gaps or space to slide beneath it. No one had fled the scene this way. She slapped a hand three times against the barrier.

The sound didn't carry.

She pressed her ear to it but heard only a rushing sound like the sea. Which made no sense whatsoever. Must be blood in her ears. But it gave her a *Truman Show*-esque sense of disorientation, as though the country she'd returned to on that long overnight flight wasn't quite the same one she'd left the previous year.

Maybe I'm dead and this is purgatory?

Stuck forever on a motorway.

Going nowhere.

She turned and the sensation popped. A normal Friday. Traffic jam. Red lights. Angry faces.

Dead driver in a black sedan.

Why would a killer attack in a place where they couldn't get away?

Some sixth sense made her look over in time to see movement beyond the central reservation close to where Daisy stood with Olly. A dull whump as a man from the southbound side collided belly first with the metal safety barrier that divided the two carriageways. Daisy yelped in shock at the impact behind her and ran forward, dragging Olly out of harm's way. The two youngsters ended up cowering beside his camper van. The intruder failed to jump the divider, but got one arm hooked over and started struggling to climb up.

Billy strode across to where the man clung to the central reservation, one elbow and now an ankle over the metal barrier. As she approached, his shoe fell off and landed at her feet.

'What are you doing?' she asked. The barrier came up to her chest. This man had found out the hard way that it was designed to be difficult to cross. Strong enough to stop the vehicles heading in one direction from crashing into the vehicles heading in the opposite direction. He hauled himself up with a huge effort and a long groan until he was perched on top of the metal rail, looking about as convincing as Putin on a horse. The leg of his grey sweatpants was now tie-dyed with black grime.

'We've lost all control over this country!' announced the spokesman from the southbound carriageway to the people of the northbound carriageway.

'Oh, aye,' muttered Pat. 'Here we go.'

'They're not telling us the half of it!' he yelled, then toppled forward so that he laid along the top of the fence in the posture of a sloth.

'You're drunk, go home!' A disembodied voice from the man's own side.

'I don't drink!' shouted the man on the fence. 'Alcohol is the opium of the masses!'

'That's religion, you dipstick,' called the omnipotent voice.

'You're going to hurt yourself, sir,' Billy spoke to him quietly. 'Don't come onto this side. This is a crime scene.'

'You're telling us nothing about it,' the man spat out. 'Someone's got to do something.'

'He's off his nut!' The voice came from between vehicles on the other side, but Billy still couldn't see who shouted.

'There's a killer!' the man on the fence bellowed. 'Round everyone up!'

'Calm down, sir.' Billy shifted so she could look him in the eye from his awkward prostrate position. He was white and in his twenties. Very pink about the gills. Gym-bunny. Solid and square as a bag of flour. No flexibility whatsoever. And dilated eyes. He was either on something or having a panic attack.

'What's your name, love?' said Billy.

'I saw something,' he whispered.

'What did you see?'

The man writhed on top of the barrier, as though unseen forces were pulling him.

'What did you see, sir?'

'It could be any one of us next!'

In fairness, he wasn't wrong. Her muscles tightened at the thought, not just that there might be a killer on the motorway, but that a mass panic sparked by this idiot could lead to a total break-down of order. But then he fell still. His face sheened over with sweat. *Heatstroke makes you vomit.* Billy stepped out of range. Instead, the man slithered off the barrier onto his own side, land-ing on his feet, still clinging to the metal panel. Billy picked up his shoe and handed it to him. They were eye to eye over the metal.

'Are you alright?' she asked.

The man took the shoe.

'Just breathe,' Billy said.

The man pouted and blew out an unsteady breath.

'What's your name?'

'Cal.'

'Cal? Cal what?'

'Calvin Barnes. We should round everybody up, keep an eye on them. We're sitting ducks. These car bombs—'

'They're far from here. It's all in hand, Cal. I'm a police officer. Sergeant Kidd, everyone calls me Billy.'

'But how—'

'If everyone stays in their cars, keeps their doors locked, we will all be safe.'

Cal flung one arm up, like a kid asking a question in class. His voice shot up high too: 'How can you say we're safe when there's a dead man? We need to find who did it before it's too late. He might pick us off one by one.'

'What do you think she's doing, pal?' Pat appeared, clasping Billy's shoulder too tight. 'She is a police officer. Did you not hear that? And she was investigating before you started playing silly buggers. Why don't you shut up, put your head between your knees, and calm the fuck down?'

It wasn't the approach Billy would have taken to bringing a man back from the brink of sanity, but by luck it seemed to work in this instance. Calvin Barnes turned on his heel, marched past three cars and a big black van to a small hatchback and got into the driver's seat. A girl sat inside, puffy-faced. When the door slammed, Billy made a hand gesture that told her to lock him inside. Which she obediently did.

Billy made a mental note to check on the pink-faced girl – and the boyfriend – later on.

On the far hard shoulder, three lads milled about in a patch of shade with hands deep in pockets.

'Keep an eye on him, would you?' Billy called out to them.

One of the lads nodded. One of them muttered *pussy*. The third giggled at nothing.

She had to get a move on. It wouldn't take much for this lot to turn vigilante.

Chapter Six

5.25 p.m.

Billy needed a cordon around the black sedan, right now. Priority – preserve the scene.

Did anyone have tape or rope?

'Don't you have traffic cones?' said Pat. 'Bollards?'

Bollards to you too.

'I'm on my way home from the airport,' Billy said. 'I don't take my stab vest on holiday. Not when I go abroad, anyway. I might in this country.'

Only Pat found that funny.

'Ah, sorry?' said Olly. 'I have climbing ropes? In the van? For the cordon?'

'That would be useful, thank you.'

Olly hustled off to be useful. He was a young man who needed a haircut – and arguably a slap because of that accent – but he did like to be useful. A cordon would help Billy maintain control. It was remarkable how reluctant people were to cross a psychological barrier; a wafer-thin strip of police tape, a cone, a laminated notice. These objects had no bulk but so much power. People submitted readily to the authority. It was weird, quite frankly.

Olly returned with two coils of thin fluorescent rope.

'Do you think someone could have thrown something from a bridge?' He pointed back down the carriageway. 'Sometimes kids chuck bricks at cars . . .'

Billy pursed her lips. She wasn't going to tell him about the skewer.

Instead, she gestured at the long clear stretch of motorway behind them. 'No bridges for a long way,' she said.

'And there's no brick in that car,' said Pat.

Billy shot her a warning look.

'Well, not a brick, then, something else?' said Olly. 'I'm just trying to think what could have happened to him.'

'Not that,' said Pat.

Billy raised her eyebrows and Pat zipped it.

'Thank you for the rope,' Billy said to Olly. She turned towards the sedan, almost slamming into Charlotte who was standing right behind her. The woman held the yellow tin of Rescue Remedy in her hand. It was an interesting approach, Billy thought, to walk about advertising your anxiety. Perhaps people pick up the subliminal clue and go easy? Perhaps Billy could try it?

Up close, the woman looked frayed, a little less expensive. Her mascara was smudged beneath one eye, like she hadn't washed her face when she got up this morning. On this muggy day, she was wearing a thin knit that covered her down to the fingertips. The wrist of one sleeve had a ragged hole that made Billy think of her grand-niece in Australia who was five and had a habit of sucking her sleeve until it was wet to the elbow.

'Hi,' said Charlotte on an out-breath. 'Thing is, my kids are with someone. I only popped out for an hour. I wondered how much longer we might be stuck here?'

'You might want to let your sitter know it'll be a while.'

'How long?'

Billy let her eyes flick to the dead man.

'Right,' said Charlotte. 'Right, okay.'

'Did you notice anything strange as the traffic stopped?' Billy asked.

'Not really, I was concentrating on the road ahead. And—' Her hand went to her fringe, pulling it down over her cheekbones.

'And . . . ?' Billy prompted.

'I was on the phone.' Her eyes were owlish, the focal point of her face, but her delicate nose was red with a few broken veins. *Drinker? Frequent tears?* 'The phone was hands free, of course.'

'Of course,' said Billy. 'So you didn't notice anything untoward as the cars stopped?'

'Untoward?'

'Did the black sedan seem out of control?'

'I didn't notice it until we got out and the driver was . . . not moving.'

Fair enough. Billy hadn't noticed Silver Fox or the little-and-large mother and daughter or Foul Mouth until they came to a halt either.

'He could have been poisoned,' said Charlotte in a rush. 'By someone. His wife. You could give someone a slow-acting poison before they go off for a drive and they'd crash and everyone would think the accident was the cause of death, not the poison. Only in this case the traffic stopped before he could crash.'

'They run blood tests,' said Billy. 'Especially in RTCs.'

Charlotte looked blank.

'Road traffic collisions.'

'Because of drink driving.' Charlotte nodded. 'Didn't think of that. It's quite hard to get away with a crime, isn't it? So you don't know when I'll be able to go?'

'I'm afraid I don't.'

'Right, I'll call my— I have to make a call.' She bustled off.

Charlotte was keen to leave. Billy noted that. But, then, weren't they all?

The woman climbed inside her Audi and locked the doors. Engine wasn't running. So no aircon. She just liked the peace. And privacy.

Anyway, this cordon. There was no point making it too wide – people had been all over the carriageway since they got out of their cars – but most of them hadn't strayed close to the black sedan. Only Pat Mackey. There was still a chance of forensics.

So the cordon needed to isolate the sedan without causing an obstacle on the hard shoulder. From her own vehicle, she grabbed her suitcase, dragged it to a spot on the road behind the sedan and tied one end of the rope to its upright handle. She ran the rope to Charlotte's SUV in the middle lane. She knocked on the window and asked if she could fix the rope to the vehicle. The woman waved a hand to agree – she was on the phone – so Billy passed the line through the door handles on the side nearest the sedan, then across the inside lane between the sedan and the mother and daughter's minivan. She wrapped the loose end around a bag containing a boxed-up bottle of gin she'd bought in duty-free, which was heavy enough to hold the rope steady. She stepped back and considered her handiwork, very much hoping no one would run over the gin.

The cordon was less of a regulation rectangle, more of a precarious parallelogram but at least the vehicle and Dead Driver were protected. People knew not to mess with her crime scene.

Her suitcase divided the space between the sedan and a sporty Range Rover that had stopped a whole car-length back. That was strange in itself. What made him stop so far away? Perhaps he had been trying to avoid running into the sedan? Perhaps it had been out of control?

The Range Rover was painted in the livery of a company called Doors & Co, in sky blue with a slogan in pillar-box red: *I'll Take You Home.*

Presumptuous.

Billy approached the driver's window, which opened, releasing a citrusy aftershave. Inside was a young guy. Well, young to Billy's eye. Mid-thirties. Estate agent brochures on the passenger seat. That made sense of the slogan. Maybe it wouldn't seem so creepy if he didn't have such finicky facial hair, like he'd cut every individual strand with a barber's blade. Made him look like a sex pest.

'Sorry to bother you,' Billy said, 'but I have to ask what kind of sociopath leaves a massive gap in a traffic jam?'

He gave a nervous chuckle. 'There was a motorbike parked there when we first stopped, but it hared off down the hard shoulder.' Lavish Brummie accent. 'Is he dead?'

'I'm Sergeant Kidd. What kind of motorbike was it that left the scene?'

'It was a Husqvarna Nuda 900,' said Brummie Sex Pest. 'Maybe he had an allergic reaction?'

'What?'

'The dead man. An allergic reaction to peanuts?'

'Peanut allergy causes anaphylactic shock or extreme swelling. I've seen it before. It's very obvious, especially on the face.'

'Bees, then?'

Billy cocked her head.

'A bee could sting you anywhere on the body,' he explained, 'but you'd still have an allergic reaction that killed you.'

Billy shrugged. Fair point. But, of course, it wasn't an allergy.

Unless he's allergic to sharp objects in the back of his neck which, if you think about it, we all are.

'I'll keep it in mind,' Billy said. 'Can we get back to this bike? It was a Husk— what?'

'Husqvarna. Courier bike.'

'Number plate?'

'Sorry, bab.' He shook his head.

'Did he dismount?'

'Nah, he was all' – he made a gesture like a fish fighting its way upstream – 'impatient, like. Revving it up. Made a right noise, hell of an engine on it.'

Billy remembered the racket. She'd had to turn up the radio to hear the news.

'Was there a passenger on the bike?'

'Nah. Just a driver.'

Could this motorcyclist have been the perpetrator?

Must have been an exceptionally fast mover ... Assassin-grade killer.

'Did you see anyone slip in or out of that black sedan?'

'Nah, I was ...' He nodded at the passenger seat.

'You were on the phone?'

'No.' He stroked his dubious facial hair. 'Well, yeah but I was going to be late for a viewing, wasn't I? It's not illegal, is it, with your engine off?'

'I'm just trying to get a sense of what went on here. With this motorbike.'

'I leant down to get the client's number from my bag and when I sat up, the bike had gone.'

'How long were you leaning down? How long was the motorcycle out of sight?'

'Not long . . . a minute?'

Was a minute long enough for the motorcyclist to jump off his bike, stab a driver in the back of the neck with a metal thing, get back on and speed off? All without being seen – by the estate agent or anyone else on this motorway?

It didn't seem feasible.

She asked the estate agent's name.

Carl Dawes.

She thanked him and stared up the carriageway. Shark-eyed cameras peered back. If only she had the motorbike's details to put through the automatic number plate recognition system. ANPR would track it within minutes. Access to the CCTV feed would answer a lot of questions too. There's no way anyone could have entered that black sedan without being seen on camera or by witnesses.

So . . .

The attacker must have been inside the black sedan.

So . . .

Maybe they were still inside the black sedan?

Chapter Seven

5.30 p.m.

Billy scissored her legs over her own cordon and slipped one hand inside the protection of a nappy bag. Better than nothing. She popped the boot of the sedan and hesitated a moment before flinging the lid up.

Empty.

Valet clean.

She slipped one finger into a stainless steel D-ring to lift the section of floor that should contain tools and a wheel. It contained tools and a wheel. She replaced it and shut the boot without touching anything else.

There was nothing on the back seat of the car or in the footwells. The only item inside the car at all, apart from the body, was a small suitcase sat on the passenger seat. It was a roll-on, the kind designed to be taken into an aeroplane cabin.

She had a sudden thought and dropped to her knees, then lay flat on the hard shoulder. But, no, there was no one hiding underneath the sedan or any of the other cars. She performed a sort of downward-dog affair to heave herself upright. Not dignified. She brushed herself down.

The exertion made her feel hot on the inside as well as the outside.

Focus.

How was this man killed . . . ?

It wasn't peanuts or bees. Maybe Olly Sims was right and the weapon had come from outside. But the glass was intact. Car had no sunroof. So, no. Or maybe Charlotte McVie had a point and something had been done to him earlier. But no, a skewer in the brain must be instant. Her first instinct had been correct; it had happened in the traffic jam. The killer was right here, in one of the hundreds of vehicles stuck on this motorway.

Look at them all, going about their lives, now put on pause.

Each one a story. Each one a suspect.

But someone knows.

Someone saw.

Can't climb over the fence to get off the motorway.

Can't run down the carriageway without being seen.

Can't climb over the central reservation without being seen.

Can't disappear into thin air. Even this thick heatwave air.

Unless literally everyone on the motorway was in on the crime, the killer must be here and they must be nearby.

She just had to find him.

The mother and daughter had left their minivan again. It seemed mum was struggling with the temperature, her face puce as she peeled her black top from her chest. Billy approached. The facial similarity of the two women was quite unsettling, twins separated by time. Except the daughter was as slender as the poisoned saplings on the verge.

'How are you, ladies?' Billy asked.

'Kerry Wells,' the mother said, as though Billy had asked *who* are you? 'And this is my daughter, Hyacinth. Are you certain he's dead?'

Billy assured them that Dead Driver was dead. The two women looked shell-shocked; Hyacinth, who must be in her twenties if she was a day, was sucking her thumb like a hungry toddler. When the mother gave the same address for them both, Billy wondered if perhaps Hyacinth didn't have both oars in the water.

Her mind chose this moment to open the rabbit hole and let thoughts of her daughter hop out. *Would she have been her own woman, like Daisy, or a mummy's girl, like this one, Hyacinth?* In all her years of masochistic fantasising, Billy had always imagined her child as carefree and content – not awkward and clingy. She was a fantasy child, after all, so who could begrudge her the positive spin?

But if you couldn't have accepted her for what she was, maybe you weren't cut out to be a mother?

Oh, jog on, foul thoughts.

Kerry was holding out a blanket. With Billy just staring dumbly at it, the mother gave it a little shake.

'You can use this to cover him,' she said.

Billy still didn't touch the blanket. 'I don't need it.'

'Or I've got a coat?'

'It won't be necessary.'

'Why can't you cover him up?' Hyacinth spoke around her thumb.

'He's just lying there,' said Kerry.

'I can't touch him until they've examined him.'

'Who?' Kerry asked.

'Forensics.'

'You could at least put this over the windscreen?' Kerry shook the blanket.

'Maybe you can stay in your car and put the blanket over your seats so you don't have to see.'

'Get in the car, Hyacinth,' said Kerry. 'Put the blanket over the seats.' The girl took the blanket but lingered behind her mother.

'Can you recall what happened as your car came to a stop?' Billy asked.

'What do you mean?' Kerry said.

Simple enough question. The mother seemed as dozy as the daughter. Or evasive.

'You were driving, right?' Billy asked.

'Who are you?' Kerry asked.

'My name's Sergeant Belinda Kidd. People call me Billy.'

'No uniform,' Hyacinth muttered.

That hadn't bothered her when she wanted Billy to climb in a hot car with a dead body and cover it with a blanket so they didn't have to see his face.

'I wasn't on duty today.' She let her eyes flick to the sedan. 'But I am now.'

'Right,' agreed Kerry.

'So was there anything unusual?'

'Nothing. We stopped in traffic. Sat there for a while. Put on the radio to get some travel news and heard about the terrorist attack. And then that woman started screaming.'

'You didn't get out of the car?'

'What?'

'When the driver of the SUV screamed, you stayed in your car.'

'You didn't need people getting in the way. And Hyacinth was upset.'

The daughter sniffed loudly in confirmation.

'Had you noticed the black sedan when the traffic was moving? Was there anything untoward about his driving? Erratic?'

'He was behind me.' Kerry pursed her lips and shook her head. 'It was all just normal. I was, you know, intent on the cars ahead, how it is when you're nose to tail.'

'Didn't hear anything?'

'Such as?'

'Voices. Tyre squeal. Anything odd.' The sound of someone being stabbed in the back of the neck with a skewer.

'We had music on. Something Hyacinth likes.'

'Dad's playlist,' muttered the daughter around her thumb. 'Britpop.'

'Retro,' said Billy.

Hyacinth shrugged.

A single shout from way behind made Billy jump and turn. Maybe thirty or more car-lengths back down the carriageway, a lorry was rocking. No, not a lorry. A horse box. Judging by the commotion, they were unloading the animals.

What the actual . . . ?

That was just what she needed: a bloody gymkhana. A child led one of the ponies onto the grass verge and it stretched out for a luxuriant pee. Made Billy think she'd been rash to drink all that coffee.

People were settling in for the long haul. Making their own rules.

Billy thanked the women for cooperating. With no uniform, no warrant card and no kit, she figured it was better to hope for goodwill than authority.

She went round behind the sedan. No skid marks on the road. The vehicle had its front passenger wheel turned onto the rumble strip. She leant her full weight on the vehicle and felt it give. The handbrake had not been engaged.

47

So. Revised assessment: it happened while the traffic was crawling along and he'd rolled to a standstill – or maybe braking was a reflex action, she wasn't sure – but he lost control of the wheel in the last seconds as he died, allowing the nose to drift towards the hard shoulder, and he hadn't been able (or alive) to pull on the handbrake.

There was something significant about the timing of the death. It felt planned. Precise.

And yet there were so many potential witnesses. Which is a crazy place to attempt a murder.

Except . . .

The terrorist attacks had caused a gridlock that trapped everyone here. Without that unexpected turn of events, all these vehicles would be long gone. Except the sedan. The sedan would have rolled onto the hard shoulder, like a regular breakdown. Drivers would have sailed past in a hurry to get home on a Friday night.

The killer hadn't accounted for terrorists. So how would he react now that the plan had gone spectacularly wrong?

Kerry aka The Mother

Toyota Sienna (2017 Model)

Can you see her, the copper, behind the black car?

I take a grunted reply as a no.

Every time I look in the rear-view mirror I see his face. Should have insisted he got covered over with the blanket. It's inhuman, leaving him there in the heat. Looking at us.

Hyacinth turns up her music.

'Don't Look Back in Anger'.

Turn it down.

Hyacinth huffs and jabs the radio. The car explodes with a woman's voice saying 'Car bomb, car bomb'. Over and over 'Car bomb'.

Turn it down!

'Car bomb!' the woman on the radio shouts back.

Hyacinth jabs again and the car goes silent.

It was too loud, that's all.

Hyacinth mutters that there's been another car bomb.

I heard. What do you want me to do about it?

She folds her arms and hunches forward to look in the wing mirror, then asks what we should do if he opens his eyes. There's a switch on my side somewhere that'll tilt her mirror, stop her from keep looking— There. It budges an inch and Hyacinth sits back with another huff. Of course, it's hard to forget he's right there, right behind us, it's perfectly normal to be curious, but even so, stop staring. And stop chewing your thumb.

It'll bleed, silly girl.

Hyacinth lets her hands drop and asks about the money. Is it because Daddy's gone?

It's exhausting, inspecting each word before it leaves my mouth, feeding Hyacinth as carefully as I did in the baby years when I bit each grape in half to stop her choking. All these years on and I'm still finding something soft enough for her to swallow. Her dad said not to worry about her, not everyone is book smart. The counsellor said 'give her room to spread her wings'. And it's true that since her dad's been gone, since I lost all grasp on reason and had – God, I hate the word, so sordid – an 'affair', there are days when I'm not sure who is the more dependent, me or her.

Even so.

Most adults can put two and two together. That's not book smart. That's just smart smart. She needs to at least try to understand.

Deep breath.

There is no money, not right now, it's been locked away for safe keeping.

Her reply is a whimper. Her voice has been like that since the baby years, not so much a cry as a bleat. On those long nights of feeding, I used to wonder – God forgive me – if the baby might accept another mother like a lamb goes to any ewe. Didn't mean it, probably most parents think that when they're bone-tired, I never thought it in the cold light of day, don't think it now either, I'm proud of her, especially recently, she's grown up a bit since her dad's been gone. Except sometimes . . .

She asks how the money can be safe if we can't get hold of it? I'm struck by my daughter's face as she bleats on about money.

I'm seeing her for the first time. The shape of her ear. Round like a chimpanzee. People say she looks like me, but—

Who even are you? What are you talking about? What do you want from me?

Imagine if I'd had a headache that one night, all those years ago, and shoved him off. Where'd I be now? Not here, that's for sure.

She's still talking about money.

Stop, silly girl! Can't stand her bleating any longer. I turn to face her. *There is something you need to know about the money.*

Chapter Eight

5.36 p.m.

Billy dialled a number she hadn't called for eighteen months. Superintendent Dominic Day. D-Day to those who respected him. Doomsday to those who didn't. Like catchy pop tunes, police nicknames didn't have to make sense so long as they stuck in the mind.

Billy called him Dom – Dominic if she was feeling sharp – because that's what she'd called him since they were thirteen and at school together. Almost four decades of over-familiarity.

Dom answered her call with a shouty, 'Billy, you're back!'

She wasn't aware he knew she'd even been away.

'How was Australia?' he asked.

'Even hotter than here.'

'How's your sister?'

'Same.'

'What, even hotter?'

She let him get the blokey banter out of his system, which was the quickest way to get to a real conversation, then she told him what was happening on the motorway.

'Blimey.' He snapped to attention. 'You really think the suspect is still there?'

'I'm telling you, they must be here. There is no way off this motorway except for running headlong down the hard shoulder.'

'No one saw them leave the scene? They couldn't have got into another car?'

'Unless every single driver on this motorway is in on it, then no, no one saw anything suspicious, no one saw anyone moving about the carriageway. It's like the suspect evaporated in the heat. Can you request the CCTV?'

'I'll put a call in. Might take a while, though. Whereabouts are you?'

'Not sure exactly, a mile or so from the Deadwall Tunnel. Hang on.' She navigated to an app on her phone, fired it up and tapped a square on the screen. 'Lady vest beard.'

'What3Words? Or have you let yourself go?'

'Should be our location, give or take, it's hard to be sure which square we're in.'

'There's always some eager beaver willing to help, send them to find a marker post as well, just in case. Sometimes the old ways are the best.'

This was true. Billy called to Olly and Daisy. She sent them along the hard shoulder to get the number from a location marker.

Down the line, Dom got interrupted by a young female voice and replied in a mild tone: 'When I say I need it now, I need it *now* now.' Billy could almost hear the constable's spine snap to attention. Dom had picked up a military bearing from his father, even though, at the tender age of nineteen, he broke the family tradition and followed Billy into the police force. Now, he gave an exaggerated groan. 'Trying to get work out of them on a Friday afternoon is like trying to get a round of drinks out of you, Billy. What did Australia have to offer that I didn't?'

'Peace and quiet.'

Dom scoffed at peace and quiet. When Billy hit the skids eighteen months ago, Dom had offered her a fresh start in a neighbouring force. He loved a fresh start. In his thirty years of service, he'd covered most police ranks – and a few specialisms – in three forces. It never rankled Billy that they'd set out together but his career sped off while hers stayed in the slow lane. She didn't see it as a matter of gender so much as personality; he'd been the same as a kid. In the school production of *Romeo and Juliet*, Dom got the part of Benvolio, the peace-maker. On the rugby pitch, he was fly half, the play-maker. On the force, he kept it up, quoting his own father who once said during a memorable speech at their school careers fair that 'a standing stone gets pissed on'. Dom couldn't leave their village quickly enough. It was Billy who had made a career out of small-town policing. Her time spent patrolling this motorway had been in training – after that, she'd gone right back to her home force and stayed put for thirty years.

'I could still swing you a job with us,' he said.

'My situation is no different now to how it was when we last spoke.'

'Which is?'

'Underwhelming. And as for us . . . I don't know what kind of dynamic I'm likely to bring to your team, probably not a helpful one.'

'Like a vindictive little sister crossed with an ex-wife.'

'I'm a month older than you, why am I the *little* sister? You're the little sister.'

'You know what I mean.'

'You just get bored easy,' she said. 'You want someone around for your entertainment.'

'And what do you want? I thought this jaunt to Australia meant you'd finally cut the apron strings.'

Billy lowered the handset and glared at his name on the screen. A moment passed in which it was acknowledged that their play fighting had got out of hand and he'd drawn blood. She put the phone back to her ear.

'Sorry, Bill. I'm just . . .' He didn't say what.

So she did. 'Disappointed in me?'

A pause. 'Disappointed *for* you, maybe.'

A small burn in her chest, like a nudge that lands on a sore spot.

'That's why I won't work with you, Dominic. They offered me retirement.'

He tutted. 'You don't have to *accept* retirement.'

'If I go back, they're going to keep me on light duties.'

'Is that why you ran off to Australia? You got the hump about light duties?'

'I've already sat behind a desk for a whole year. It was boring and I got fat. I was cleared of the misconduct and I thought I'd be allowed back out there, but—'

'Light duties is not a punishment. It's for your benefit, so you don't run straight into another RTC that sets you back. It's meant for you to get your head straight. If they think you need more time—'

'You don't know anything about my head.'

'I bloody do because we talked about it. After the scooter thing. You were in a right state. I drove halfway across the country in the middle of the night worrying about you, and then all the way back home again even more worried about you, and when I got in I read the police logs and I could see there was never going to be grounds for misconduct, but the case had to be heard – that's how it works – and I know these things take ages to go through the system and they drive you mad waiting, but in the end you got cleared. Justice prevailed. So that's the end of it—'

'The end is that a boy died,' she said. God, Dom could dance to the tune of his own voice sometimes.

'It happens, Billy. You had a bad week—'

'I had a terrible week.'

'You had a traumatic week that led to a bad call.'

'You said it was a bad week. Now it's a bad call?'

A hard sigh rattled the line. 'Go back to work, do your light duties and get up to speed. On my team or someone else's team, I don't care, but just do what needs to be done. Yeah?'

Billy had tuned out his pep talk. In her peripheral vision, she'd become aware of figures rushing towards a car. The VW Golf belonging to Pat Mackey. The nurse was sat in the driver's seat with the door open and radio playing. Nigel Heathcote had one elbow planted on her door. Charlotte McVie listened with a hand clamped over her mouth.

'What's going on?' Billy said. 'Something on the news?'

The clatter of a keyboard down the line. 'Another car bomb. Went off at precisely 5.30 p.m., this time at a bus station.'

'That's three in total?' Billy said. 'All transport hubs.'

'Exactly thirty minutes between each one. Four thirty at the train station, five o'clock in the Deadwall Tunnel, now five thirty at this bus station.'

'Why do that?'

'Because they can. Shock and awe. Keep the boys in blue busy.' Dom had done a stint in counter-terrorism. Lucky this incident wasn't on his patch or he'd be hands-on-deck. Probably wished he was. 'They've got us round their little finger now, everyone wondering where the next bomb might go off.'

Along the carriageway, the freakish May sun glinted off hundreds of windscreens.

'I've got a dead driver and the potential for mass panic,' she said.

'There you go, then – just what you need, out from behind the desk.'

She shrugged off his gallows humour. 'Can I give you some number plates? I want an ID for the dead driver. I want to know if any of the other vehicles have a connection to him.'

'Text me the numbers. Anything else you need?'

'I need to get home.'

'I've been thinking about moving back too,' said Dom. A beat. 'You know I'm separated?'

Billy's stomach flipped. Ridiculously.

'That's news,' she said.

'You hadn't heard?'

She'd kept her head down in Australia. Not one for social media at the best of times, she hadn't been in regular contact with anyone who might have relayed this titbit. But she was surprised someone hadn't gossiped. Enough people knew the ancient history of Billy and Dom. There were even those who tried to insinuate there was *recent* history; D-Day had been spotted comforting Billy outside in the vehicle yard that night when she'd summoned him after the worst day of her career. The scooter thing. One of the young PCs sneaked outside to vape and spotted them. Spread a rumour that they'd been entwined 'like a pair of old swans'. Cheeky little moo. As if Billy had the flexibility to entwine.

She cleared her throat. 'What happened?'

'She left me. I think we're getting a divorce.'

A smidgen of disappointment that it was the wife who left. The wife was something in the civil service. Actual Downing Street.

Wait . . .

When Dom asked if she'd heard about his separation . . . ?

Did he think she'd come home from Australia because she'd heard he was getting divorced? And phoned him as soon as she landed at the airport? Was he that cocky?

Yes.

Or was it wishful thinking on his part? Did he want her to call?

'What do you mean you *think* you're getting a divorce?' she said.

'She says it's a trial separation, but she does that. Lays the groundwork. Last year, she said she was thinking about going to Iceland with her book club – turned out she'd already paid the deposit. So if she says she's thinking about a divorce then we're as good as divorced. Maybe she's got a new bloke.'

'Maybe she's trying to get your attention?'

'She says I haven't fought for her.'

'Have you?'

'She does high-intensity training, I don't think I could take her.' Deflection.

'Now is not the time,' Billy said. 'But later, yeah? I'll want to know if you're alright.'

'I'm not living on frozen chips, if that's what you mean. It's all good. I bought a stand-up paddleboard.'

'Have you used it?'

'Obviously not. But it shows willing, doesn't it? The purchase of a stand-up paddleboard is a sign of optimism. I'm getting an air fryer next.'

Billy had got quite good on her sister's paddleboard in Australia. It wouldn't be as nice here, but there was a lake about ten miles from her place where they could go. Full of goose shit, obviously, but it should be fine so long as they didn't fall in. She looked fetching in a wet suit, once she got into the thing. Getting out would be undignified—

Focus.

'I should get on with it.' She glanced at the people gathered around Pat's car.

'Come back to work, Bill.' He sounded more impassioned than when he'd spoken about his divorce. 'You don't deserve to be put out to pasture.'

Never mind pasture, she deserved the knacker's yard.

She typed the vehicle registration numbers into WhatsApp while she talked.

'Just run the plates I'm sending now and tell me who's in these cars.'

He agreed, but sounded piqued.

'You need to be careful out there,' he said.

'I'm going to frisk this dead man.'

'If you think the suspect's still there, you should back off.'

'I'm going to find out who this dead driver is.'

'Billy—'

She cut the call.

Chapter Nine

5.44 p.m.

Over the years, there had been countless crime scenes but this was by far Billy's strangest. Not the most harrowing, not yet, but the strangest. Potentially, it was also the most dangerous. A suspect here somewhere. A killer.

That was easy to forget in this strange atmosphere that felt detached from reality. Listless heat. Muted sound. Lives paused. It created a slippery sense of lawlessness, somewhere between a bank holiday and a hostage situation.

Billy wanted to know more about the dead man.

She hoped the cabin bag would prevent her from having to frisk him.

Billy put two of Charlotte's nappy bags over her hands before opening the passenger door. A hot blast of trapped air carried a smell that made her step back and give it a moment. Not putre-faction, not yet, but a concentration of bodily odours. Something like offal. When she finally leant inside the cabin of the vehicle, she had a good look around and took photos. Nothing out of place, nothing in the car except the small suitcase. Dead Driver remained slumped. There was more blood on the collar of his white shirt now, enough that she could taste old pennies. She removed the suitcase and closed the door on him.

It was an expensive roll-on. Metal and ribbed, space age. It had a built-in lock, but the bolts clicked open. Inside, the bare minimum. Shirts, two smart and two casual. Trousers. A wash bag with nothing odd inside.

Don't make me fish in your cold dead pockets.

Give me something with a name.

In a zipped section, a wallet and a passport. An American passport.

That's going to make it complicated.

At least the passport gave her a name: Chad McClusky. The photo matched the dead man. The wallet contained a good deal of cash, crisp notes in high denominations. About £500 in fifties. About the same again in twenties. Who carries fifty-pound notes? You can hardly spend them these days due to counterfeiting. Answer: someone who's just bought their cash from a bureau de change.

There was also a credit card in the same name. Capital One Visa card. The valid dates showed it was new, its expiry not for another five years.

As Billy considered the ramifications of her victim being an American national, her gaze wandered. On the other side of the motorway was a big black van. It was parked on the lane beyond the central reservation, almost directly opposite Heathcote's BMW and, of course, facing in the other direction.

A black van with tinted windows. Printed on the side: Hybrid Welfare Unit.

What did that mean? What was hybrid, the welfare or the unit? And a hybrid of what? It sounded a bit . . . made up. A bit CIA. It was the sort of surveillance van you'd get in *The X-Files* or any show involving the FBI, where it would sit in plain sight and look so blandly, boringly official – not to mention intimidating –

that no one questioned it. It looked – incongruously, in this very mundane English setting – American. Like Dead Driver.

And its driver hadn't got out. Most people were venturing out of their cars now, but not Van Man. Was that strange? It felt strange, but then everything about this situation felt strange, and Billy's sister in Australia had been partial to the kind of box sets that fed you conspiracy theories. They'd watched a lot. So she had to stop herself going off on flights of fancy.

'Sergeant Kidd!'

Daisy came loping down the hard shoulder with Olly. The long sleeves of his shirt were rucked up above his biceps and he had sweat patches. She remained cool in her mermaid dress. They were disgustingly healthy and attractive young people. Faces smooth as pancakes. Their arms bumped together as they walked.

'Do you two know each other?' Billy asked when they arrived.

Daisy widened her eyes and shook her head. 'We met—'

'—literally right here,' finished Olly, his Byronic curls bouncing with the extraordinariness of it all.

'Only you seem . . .' Billy had been going to say *familiar* – her mind skipped back to the Rescue Remedy pastille slipped from one mouth to the other – but it was none of her business how familiar they were. 'Never mind. Did you get it?'

'I photographed it for you,' said Daisy. She flipped her phone to show the image of a wonky white post stuck in the ground. Not much to look at, but its unique number would tell the police exactly where they were located.

Billy typed the number into her phone and texted it to Dom.

'Oh, hello!' said Olly to a sandy dog who had arrived unnoticed and introduced himself by pressing a slobbery muzzle into the lad's hand. The dog was dragging a lead. 'Where did you come from, boy?'

'Is he with those horse people?' Billy peered up the hard shoulder at the horse box, whose occupants were grazing placidly on the verge. 'Can you take him up there? But before you go' – she pointed at a bone-shaped container fixed to the end of the dog's lead – 'grab the poo bags, I need them.'

Olly just managed to unclip the plastic container before the dog pricked his ears and ran, tugging the lead out of his hand.

'He's away!' said Daisy.

The dog loped off down the carriageway in the opposite direction from the horse people. Billy muttered under her breath. Next thing, someone would be roaming around looking for their lost dog. Why couldn't people just sit quietly and wait it out? You'd think the centre could hold for more than an hour.

'Make sure the dog's alright, would you?' she asked the couple. They trotted off. At least she'd got some more plastic bags.

Right, focus.

Bag the evidence. The poo bags were thinner than the nappy bags (alarmingly). They seemed to be made of some strange soft eco-material that was almost see-through, which would be better for inspecting the evidence once it was sealed. She crouched down over the suitcase, dropped the wallet and passport into separate bags and tied them up. She took the suitcase to her own car to lock safely in the boot where no one could get to it.

The poo/evidence bags were hardly tamper proof, but they were better than nothing. She tore paper from her rental agreement, borrowed a hairy-nibbed biro that Pat salvaged from the bottom of her handbag, and wrote evidence labels. Date, time, item number and signature, plus a description of where the evidence was found. Then she put the bags in her glove box and locked her vehicle.

She flapped the bottom of her shirt to let the air circulate. The activity had left her 'glowing'.

What next?

Chad McClusky. American. She texted the new information to Dom. Wished she had a radio. His car must be a rental. No one but a serial killer keeps a boot that clean.

Her phone rang at once.

'He picked the hire car up from the airport and I just spoke to them,' he said without preamble. There was a delay on the line and they both spoke at the same time like awkward teenagers, until he said, 'You go first.'

'The deceased is called Chad McClusky,' she said. 'I found his wallet and he matches the photo on his driver's licence and passport.'

Dom asked what she was doing with evidence.

'I'm having to make do and mend. I'll take photos of everything in situ. I've got nappy bags on my hands and dog poo bags if there's any evidence I have to touch. But my aim is to leave well alone until Major Crimes gets here.'

'It'll be a while. They've put out a mutual aid request to surrounding forces. They don't need my crews yet, we're too far away, but it might come to that. So you're going to be single crew for the foreseeable, Billy, you alright with that?'

'Don't *kid gloves* me, Dominic. I've done nothing but single crew these past years. We're always short staffed.'

'This is a bit different, though, isn't it? You need to be careful.'

'I'm on light duties, remember? I am careful.' Billy opened her boot and checked the airport code letters printed on the luggage tag of the case. 'Just tell me which airport is IAD?'

'It's Washington International – don't know how to pronounce it – Dooles, Dull-ess? Anyway, the woman at the car hire company was able to tell me that he landed here from Washington around midday today. Took a while to clear immigration, had to call the car hire office to warn them he'd be late, but then he turned up and everything was normal. Got the keys, took one of those buses to the long-stay car park, and presumably went on his merry way. We've got a request in for the airport's CCTV and I've got someone on the ANPR.'

Billy nodded. The system would come back with hits right away. They'd know exactly where the hire car had been, so long as the driver had stuck to major roads.

'Talking of CCTV, how soon can you get the footage from this stretch of road?'

'They said it might take a while – there's a priority on tracing car bomb vehicles.'

'If they're going off every half-hour, how long until the next one?' The heat made her brain slow. She couldn't get a fix on the numbers, the times.

'Less than fifteen minutes. But I don't think you need to worry.'

'How can you be sure?'

'The locations. The first one went off in the north of the city at the train station. Deadwall Tunnel is in the east. This latest one at a bus station is in the south. It's like they're working round the compass points. Going clockwise.'

'But what if ... ?' Her gaze skittered over the surrounding cars. A less rational person might feel a prickle of panic.

'The pattern suggests it'll be on the western side.'

'Not reassuring enough, Dom.'

'There's a vanishingly remote chance that one would be right where you are. There are 2.6 million passenger cars in London.'

'I'm not in bloody London.'

'The point is there's a lot of traffic in cities, so those are remote odds.'

'The point is, if I'm worried, then other people are going to be worried too.'

'So you're going to use your considerable influence to keep them calm, Sergeant Billy the Kidd.'

She'd be a lot calmer if she knew where this murderer was lurking.

'What do we know about the dead man?' she asked. 'Chad McClusky?'

An electronic ping sounded down the line. His computer, by the sounds of it.

'Hang on a minute.' *Click, click, click* of a mouse. Then: 'Don't look now, but this BMW . . . That's the one in the outside lane, right?'

'Mm hmm.' Billy set off on a little stroll along the hard shoulder in the direction the traffic should be moving. From that distance, she could let her gaze wander off leash. It skittered over the cars, past Kerry and Hyacinth in their minivan, past her own hire car, to the Beamer where Heathcote sat in the driver's seat.

'Right,' said Dom. 'The BMW is registered to the King's Arms public house in Howle Green.'

So the silver fox was a pub landlord. He fitted the bill.

'What's the significance?' she asked.

'Chad McClusky gave an address in Howle Green to the rental company. An Airbnb. It's a small village, so that's a coincidence. It's close to where you are, couple of miles east.'

Heathcote had Ray-Bans on and was singing to himself. He looked like a Bryan Ferry tribute act. He looked like he had not a care in the world. Doth he protest too much?

'That could be why the killer hasn't left the scene,' said Billy. 'He's stuck in the outside lane, can't get out of the queue. And if he abandoned his vehicle it would only draw attention to himself. So he's decided to hunker down.'

'The safest course of action for you now would be to hang back and keep eyes on him until another crew arrives.'

Billy muttered something.

'The line is breaking up, did you say "roger that" or "bugger that"?'

'What d'you think? I'm going to talk to him.'

Heathcote aka Silver Fox

BMW 3 Series Mk6 (2015 model)

The whole steering column vibrates when I drum the wheel with flat palms. *Bam, bam, bam.* Three hard notes to end the song. Banging tune. A track should come to a hard end, not fade out like no one knew how to make it stop.

Then the DJ spoils it by talking. I flick the volume down until he's a whimper.

Check my mirrors.

Nothing doing.

All quiet on the western front.

I'll give my princess a quick call, then.

It rings twice and she sounds happy when she answers. She can't talk, though, she's at work, she's on in a sec.

Don't have a minute for your old dad?

She's got literally one minute then she's on.

I'm going nuts here, stuck in traffic.

She knows. She tells me what I've already heard on the news. I don't tell her what else has just happened here, it's not the sort of thing you share over the phone.

She's got to go, but asks if I'm alright.

Just letting you know I love you.

There's a pause then she asks if I'm in hospital.

Why would I be in hospital? I'm on my way home. Been a long day. I'll tell you about it when I see you.

When I break her bloody heart, I mean.

She's saying she's got to go but what's all this about, what's happened?

It's alright. I'm alright. I just love you that's all.

She says give over, you're worrying me now.

Don't be worried. Go and shine. I'll be listening.

She rings off.

Hope she doesn't get suss and start looking into it. Should have kept schtum.

I tap my palms against the wheel to find a rhythm – give me a minute, it's been a while – then I'm drumming the opening to 'Highway To Hell'. That double kick on the bass. Or the brake pedal. So good! I miss drumming. Might get back into it when I'm a man of leisure. The exertion, the focus, the release. Don't miss the palaver of packing the kit into the van at the end of the night, mind you, the other guys already at the bar with the best of the girls. I'll get some roadies. Those were happy days, though, happy days.

Happy days are here again.

Judas Priest, I hope so.

Could do with a bit of luck.

Check the mirrors.

Still nothing doing.

Along the carriageway, red lights shimmer in a migraine aura. Not now. That's all I need. I blink the dizziness away.

We can't be stuck here much longer, surely?

The shimmer returns. The heat gets to my head.

Worse when I'm tense, I need to take something.

Check inside the door pocket. Balled-up receipts, spongy old tissues. And a blister pack. Come on, must be one left. My thumb finds the smooth surface of an unpopped pill. Bursts the foil with a nail edge, frees the tablet. I can hawk enough to swallow. There you go. Down the hatch.

Happy days are here again.

Not supposed to take the pills while driving, but I never get drowsy and, in any case, drowsy is better than a migraine. Need to keep a clear head today of all days.

Broken hearts later. Won't be pretty, but once that's done—

Smooth sailing. This stress ends. No more migraines.

So long as we get off this bloody motorway.

Check the mirrors—

And that policewoman is walking up the hard shoulder.

The fuck does she want?

That's it, love, jog on.

Walk on by.

Oh, it's quarter to five. Nearly missed my girl's big moment. Should have tuned in more often to hear her, the raspy little voice all grown up, but I can't stand that DJ, thank God she left him, he was beneath her, that fake laugh – u-huh, u-huh – bless her cotton socks, she's got worse taste in men than her mother. God rest her bones 'n all.

The jingle plays. Soft-rock riff with car horns bouncing side to side. Cheesy. I should have written her something myself. And there she is for the last time.

I'm Amy Heathcote with your traffic and travel. Lots to report on the roads this evening, a number of major incidents causing chaos throughout the city. The Deadwall Tunnel remains closed in both directions—

So I'm stuck here for the foreseeable. The shimmer's back, blurring my edges. Should have taken the pill quicker. And the copper's on the move again.

The fuck does she want?

Incoming.

Chapter Ten

5.49 p.m.

Billy crossed the carriageway to the BMW.

'Mr Heathcote,' she said, with a knuckle rap on the driver's window. 'Nigel Heathcote?'

He opened the door and stretched one long leg onto the tarmac. Then followed it with the lanky rest of him. The air conditioning splashed over Billy's face as he slammed the door and reclined against it, arms folding over his chest. Up close he was a fine-boned man with a neat quiff of grey hair above a woollen hoodie with a Japanese logo. He was older than his youthful clothing suggested, eyes etched and shaded. Too many late nights behind the bar? The silver fox resembled some actor whose name Billy couldn't summon, but it was one of the ones who played seasoned detectives that sleep with girl officers half their age.

Focus, Belinda.

'Could I ask you a question, sir?'

'Sorry, I'm spoken for.'

What?

Oh.

Dickhead.

'Are you the landlord of the King's Arms in Howle Green?'

'For my sins.'

'The deceased man in that car is called Chad McClusky,' she said.

He pushed out his bottom lip and shook his head.

'Mr McClusky was heading to Howle Green straight from the airport.'

His eyes went wide. Not an actor after all; he'd overdone the surprise.

'He was headed to an Airbnb about half a mile from your pub.'

'Don't know it,' said Heathcote.

'Is Howle Green a big place?'

'Just a village. Cut in half by a bypass. Which is killing us. Less traffic means less business. They say it's progress, helps people get into the city quicker, but there you go.'

'So it's a small village?'

'Yeah.'

'I remember when the pub was the heart of the community,' Billy offered. 'A pub landlord would know everyone.'

'The good old days. Now they keep themselves to themselves.'

'Who do?'

'Everyone. They'd rather sit in and watch Netflix.'

'What about the Airbnb? Much call for that in a small village?'

Heathcote blew out his cheeks. 'We're not a tourist village. When visitors come, they're on business. Don't spend much time in the pub, you know? We're more for the locals.'

'What kind of business brings visitors to a small place like Howle Green?'

Heathcote spat out a sharp breath. 'We're the last to know. You'd do better to ask the MOD.'

Ministry of Defence?

And then Howle Green rang a bell.

RAF Howle, the air strip used as an American military base in the UK.

Another American connection.

Who are you, Chad McClusky?

The airbase rang a bell because it had been in the news a while back. A diplomatic incident. An American who worked at RAF Howle had ridden a bicycle at high speed through a red traffic light. He collided with a local man at the crossing. Before the victim had even died of his injuries, the American – rumoured to be a spy – had fled the country under diplomatic immunity. The US government failed to extradite him. It resulted in the British prime minister and the American president refusing to shake hands at a NATO summit. First time in history that had happened.

The incident had rumbled on for months but with no fresh developments, it fell out of the news cycle. But the name of RAF Howle would forever be synonymous in the public mind with injustice.

'What was his name again?' Billy asked. She had her phone out, found the Wikipedia page, but the data was moving about as fast as the traffic. With everyone making calls, streaming live news, sending messages, the system must be stretched to capacity. 'The diplomat, what was his name?'

'Diplomat? You mean killer?' said Heathcote. 'Who got away with murder?'

Billy lowered her phone and watched him. Seemed like he had a bite. She arranged her face into a sympathetic smile. She knew how long it could take a community to get over a death.

'The American. What was his name?'

'Lincoln Quick.'

Of course. Lincoln Quick. Not Chad McClusky.

'Have you ever heard the name Chad McClusky, Mr Heathcote?' Billy scrunched up her face like it was a puzzling mystery they were contemplating.

'Nup. I didn't like to say before, but . . .' he said, then shook his head and looked away.

'What?'

He huffed. 'I didn't like to say because a man's dead over there, and it doesn't do to speak ill of the dead, but the Americans don't come in my pub. If any one of those bastards caught fire, I wouldn't throw him the contents of a spittoon. So, no, I don't know this bloke or anything about his shady business. If he's a spook, then he's probably been bumped off by one of his own. That was more than one question and I'm getting hot in my cashmere, so are we done?'

'We're done whenever you like, Mr Heathcote.'

He got into the driver's seat and slammed the door. The powerful car purred to life and for a confused second Billy thought he was going to drive off, but of course he couldn't because he'd have to ram other vehicles out of the way to get anywhere. He settled back in the driver's seat and closed his eyes. She felt the steady heartbeat of his music through a throaty stereo. And he must have all that sweet, sweet aircon.

She checked her watch. Ten minutes until another car bomb might – or might not – go off.

As she took the few steps back to her car, a fizzy pressure in her undercarriage told her she might go off herself. The scruffy trees did not provide enough cover and she didn't want to leave the scene to look for a quieter spot. She loosened her belt a notch and that helped.

Okay. If another car bomb went off at six, people were going to notice the timing – one every half an hour – and the atmosphere here was likely to get a lot more agitated. So she had ten minutes to make progress. Gather some facts.

What next?

Carl Dawes, the estate agent with the sex pest beard, was still inside his gaudy Range Rover.

Literally the only movement on this motorway since the murder happened was the motorbike that supposedly left the scene from the space in front of him. She wanted to know more about that.

As she walked across the carriageway, she noticed that his sky-blue tie was the same colour as the logo of the estate agency. Stickler for details was our Carl.

'This bike . . .' Billy said, all chatty as he buzzed down the window.

'Husqvarna Nuda 900.'

'What colour was it?'

'Red.'

'Driver's helmet?'

Carl hummed. 'Just black. Nothing distinctive.'

'What made you think he was a courier?'

'Those road bikes normally are . . . Oh, and because he had a black plastic box strapped to the seat. For parcels, you know?'

'Was he wearing leathers, high vis?'

'No, actually, camo trousers. I noticed that. Thought he'd skin himself if he came off.'

That gave Billy pause. 'No leathers?'

'No, bab.'

Couriers were normally prepared for all eventualities so they could take a job on the fly. No leathers meant the biker wasn't planning to go far. Perhaps only far enough to carry out one very specific job . . .

Or perhaps he'd nicked the bike and was having a joy ride.

She texted it to Dom, asking him to check stolen vehicles and CCTV for sightings of the bike leaving the scene. Her phone rang at once.

'You want to hear a funny thing?' he said.

'Tickle me.'

'That stretch of motorway isn't covered by CCTV. It's a dead mile.' Dom paused to let the news land. 'So I can't tell you anything about that motorbike or any other vehicle. The surveillance cameras are switched off at the moment. System is being upgraded.'

'That's convenient.'

'You are standing on one of the only unmonitored stretches of motorway in the country.'

Billy tried to let the implications sink in, but her mind was too choppy.

'Don't laugh, but—'

He did laugh, but only once. 'Go on . . .'

'It all seems so improbable. Could this be an accident? This metal thing in the back of his neck, could it be something from his seat? I'm thinking if he hit the brakes too hard, his head whipped forward and back, impaling him on something sticking out of the headrest?'

'What's the metal thing look like?'

'It's black. About the thickness of a kebab skewer. I can't think what would be inside the seat like that?'

'A spring, like in a mattress?'

'You wouldn't get that in a headrest, though?'

'Or a seat,' said Dom. 'Not since the sixties.'

'Okay, it's not an accident,' said Billy. 'I just had to get that off my chest.'

'Always worth spitballing.'

'And there's no CCTV on this road?'

'None.'

'Then we know nothing about the motorbike and—' She was distracted by more shouts further up the carriageway.

What now?

It came from a pod of people who had gathered about twenty car-lengths back. She ignored it and carried on talking to Dom. 'Have you got anyone who could trawl social media, see if this Chad McClusky has British connections? Maybe find out why he's here?'

'Roger that.'

She started explaining about the conversation with the pub landlord until she became aware that she was speaking into a void.

'Dom?'

Nothing.

'Dominic?'

No.

Billy looked at her phone. She had signal, only 3G but something. It should work for a straightforward call.

A text came through. From Dom.

Cut off. Sit tight and be careful

She tried to call him back but got nothing, just the ghostly sound of the ether. She pocketed the phone. The stress of working single crew was nothing new to her, but she usually had a radio, the reassurance of a reliable connection to base. Now, despite being surrounded by people – or perhaps because she was surrounded by people – Billy felt exposed and alone. And the commotion up the carriageway was building. Olly must have wandered up there, as he was jogging back towards her now. There was something afoot. When he arrived, the hair on his nape was soaked with sweat and stuck to his neck in the shape of arrows.

He brought news that someone had been seen running off. They'd left behind a car. An old banger.

A cold tickle ran down Billy's spine like a jellyfish she'd brushed up against in the ocean one day off Perth. Even though it felt like nothing at the time, boy, had she regretted it later. Hadn't reacted fast enough. She wouldn't do that again.

The commotion was twenty cars back. How could someone running off up there have anything to do with a killing down here? How would they even have got from here to there without being seen?

'It was a woman. And the thing is—' Olly stopped and pulled his mouth into a tight grimace. 'They're worried about car bombs.'

Of course. An abandoned vehicle. The attacks in the city.

The commotion didn't have anything to do with her dead driver.

It was the other thing.

'The radio says a car bomb has gone off every thirty minutes. On a schedule. So the next one would be . . .' He checked his watch.

Billy already knew. 'Six minutes from now.'

Chapter Eleven

5.54 p.m.

'Stay here and watch this cordon around the dead driver,' Billy told Olly. 'If anyone approaches the black sedan, ask them to stop. Tell me about any movements when I get back. No one touches the sedan.'

'Obviously.'

'Get your friend to help,' Billy gestured at Daisy. 'She seems switched on.'

Olly smiled as though she'd paid him the compliment.

Billy jogged off against the direction of traffic. Her jeans tugged at her limbs, denim stuck to clammy flesh. She reached the knot of people about twenty car-lengths back.

'Are you the copper?' a gammon-headed man asked. He must have slathered sun cream on because the white sweat on his forehead looked like beads of fat. 'We're worried about this vehicle, and the lad mentioned you're police.' Gammon pulled up to his full height with the self-importance of being spokesman. 'It's been abandoned, and we thought— *Are* you a copper? Shouldn't you be in uniform?'

'I was off duty, hence the civvies.' Billy looked down at herself, mentally agreeing that it was an unimpressive sight. 'And, just as an FYI, a police officer is not required to be in uniform. As soon as I exercise duties, I'm on duty. So here I am. Sergeant

Kidd. It would be great if everyone could move away from this vehicle, just to be on the safe side.'

Billy saw that some people had already taken themselves off to the grass verge. A few bystanders hurried away now, but most hung close. Didn't take common sense advice from a police officer whose face was as wrinkled as her jeans. There were about ten people gathered around the car which was parked in the middle lane. It was ready for the scrap yard.

'It's not like there's an obvious device, but, you know . . .' said Gammon.

Billy did know. She checked out the vehicle without touching it. 'Its number plates are old, they match the vintage of the car. So doesn't look like it's been fitted with new plates to avoid recognition by ANPR cameras. No obscured or altered characters on the plates either. The vehicle isn't slung low on the rear axle like it might be loaded with heavy explosives. It isn't smoking or actively on fire. These are all positive signs.'

'You work in anti-terrorism?' asked Gammon.

'Thirty years in the force. I started in the days of the IRA. We all knew about car bombs back then.'

'But you've got to admit that an abandoned car seems a bit suss,' Gammon pressed. 'Woman driver walked off as soon as the traffic stopped. Calm as you like.'

A woman with yellow hair marched up and joined the conversation. Her wide-legged stance could withstand a gale. 'That woman was too calm,' she muttered. Billy's teenage niece would say she had *Karen hair*.

'She walked, not ran?' said Billy.

'Walked,' said Gammon.

'Comfort break?' Billy offered.

'Been gone ages,' said Gammon. 'Since right after the traffic stopped.'

'How did she know the traffic wouldn't start moving again?' asked Karen.

In fairness, abandoning a car on the motorway was odd.

'So how long ago did she leave?' Billy asked.

'What time did the traffic stop?' said Gammon. 'Must be coming up to an hour.'

Billy flicked a glance at her watch.

Five minutes until the next car bomb.

She needed people to move fifteen metres away, that was the minimum. Four to five car-lengths. Maybe more due to the likelihood of flying glass.

'You know about these bombs going off every half-hour?' said Karen, clutching Billy's bare forearm with a dank palm. Blocky jeans and no waist added to the effect of her being a Lego figurine.

A murmur ran through the group. A grey-haired woman hurried off. A bloke in painter's trousers gave an apologetic shrug and went too.

'The woman who abandoned the car, she was wearing—' Karen leant in to Billy's personal space and ran one hand around her face in a manner that was intended to communicate something significant.

'What?' Billy kept her face slack.

The woman waved her hand around her face where a headscarf might go. Her eyes were puddle grey. 'One of them – you know – head things.'

'A balaclava?' Billy enjoyed goading casual racists. If you want to say it, say it. Don't drop hints and hope someone will say it for you. When Billy failed to do the dirty work of saying it for her, Karen roughly released her arm, leaving behind a sharp sting of nails. Billy pointedly broke eye contact and asked them once again to move back. To the grass verge. Fifteen metres away. She explained about the flying glass.

Karen turned one shoulder, shutting Billy out of the group. 'We should move the car. If we all got together we could push it away.'

'Don't touch a suspect vehicle.' Billy spoke in an even voice. Karen's shoulder twitched like she was shrugging off a fly, but Billy persisted. 'If there's a suspicious object, the advice is to move away.'

'Yeah,' said Gammon in a drawn-out tone that might be appropriate for deliberating whether to supersize a McDonald's, but not for a matter of life and death. 'It might go off, I suppose, if we touch it.'

Of course it might go off, you numpty.

She'd come over here thinking she might struggle to keep them calm. Instead, she was struggling to wake them up. She worked her way around a couple of orbiting bystanders, appealing directly to them to move back to a safe distance. The grass verge. A man wearing a backpack on his chest left. A woman in a straw hat went too.

Karen was still going on about how her car was parked next to the car bomb, and she couldn't afford to lose her vehicle, so they should move before it's too late.

You must have insurance, woman?

Billy glanced at her watch. Four minutes to go.

Only Karen, Gammon and a middle-aged black guy in retro glasses lingered.

Billy got down on her hands and knees to peer under the abandoned car. She didn't know what she was looking for, to be honest. Wires. Gas canisters. A flashing red light.

A stick of dynamite with ACME stamped on it.

Car bombs were not her field of expertise. Her fields of expertise were teenagers who crashed their parents' cars into trees, the theft of gardening equipment and small dogs, and the retrieval of silver vials of nitrous oxide that gas-sniffers left scattered about her town. There were hardly any terrorists on her usual patch. She only knew the basics, and that dated back to another era.

At the start of her career, Irish Republicans had used car bombs to generate a background hum of fear. The thought of them had been terrifying, but only because no one had heard of al-Qaeda yet. They upped the game with suicide bombers. Then she understood what terror meant; it meant facing a killer who didn't care if they lived or died. There was something inhuman and inconceivable about the willingness to die for a cause. Never mind the laws of the land, what terrified her was a person who had stopped obeying the laws of human nature.

It was too soon to know what brand these terrorists in the city might turn out to be. She could only apply the rules about car bombs that she'd learned back in the day.

So . . . her training said to look for anything HOT. The H was for Hidden, which meant disproportionately heavy packages hidden inside everyday objects. She'd already decided the car didn't look weighed down. The O stood for Obvious. Well, this whole situation was *obviously* suspect. And the T was Typical. Look out for anything *not* typical. The underneath of this car was typical for an old banger, rusty and oily. So she only had an O. She took a big sniff. Some explosives smell like almonds. She smelt only fumes and herself.

Billy got up, stiff-limbed in sticky jeans. She peered inside the vehicle. Car bombs could be triggered by door handles, although that was old school. A mobile phone would be the detonator of choice these days.

And that's when she spotted one inside the car. A simple Nokia burner-style phone in the centre console, on charge.

Bugger it.

Who leaves their phone behind? It wouldn't even be charging with the engine off. That was not Typical. Now she had an O and a T. And only three minutes.

She pointed at the drivers.

'I want you all to move back. Now!'

Retro Man spoke for the first time. 'Maybe we should evacuate the whole motorway?'

'Let's talk about it over there.' She spread both arms wide, shepherding them away.

'What about my car?' Karen's voice cinched tighter. She pointed at Gammon. 'If you pull yours onto the hard shoulder, we can all shunt forward and make room.'

'Please, madam, move away. The hard shoulder is for emergency vehicles only, it's not a car park. Let's go.' She ushered them along for two car-lengths.

Retro Man started again. 'Why not evacuate? First there's this car and then the dead man?'

'There's nowhere to evacuate to, sir,' Billy said. 'It's a good couple of miles to the exit.'

'That's got to be safer than here?'

'Any busy public location would be an attractive target to a terrorist, so I'd say it's much of a muchness. We're probably safer here because no one can drive up with a car bomb. Not with the road blocked. Shall we keep moving?' Billy succeeded in shepherding them another few metres. It was like the three of them were shackled together and she couldn't get one to proceed unless the other two were moving too.

'We could all walk down the hard shoulder and off the motorway,' said Retro Man. The guy was a stuck record.

'And go where?' said Gammon. 'She's right. We're better off waiting it out.'

'At a safe distance,' said Billy.

'Like sitting ducks?' said Karen.

'If you think about it, nowhere's really safe,' said Gammon philosophically.

'The grass verge is safe. Distance and protective cover are the best way to avoid injury.' Billy kept her voice soothing. 'Keep moving. Another few metres.' It was like getting children to eat their vegetables. 'Two more car-lengths. That's it. Away from any flying glass.'

Away from the exploding bloody car.

Two minutes.

They crossed the hard shoulder onto the grass verge and she almost cheered. Billy got out her phone as they finally made progress. She tried Dom's station number, getting an unfamiliar flatline tone. She texted him the number plate of the abandoned car. Not sure if it had gone through or not, she left it to keep trying.

'*Hey!*'

The voice came from behind her. Across the carriageway on the other side of the central reservation, a young man had his arms and shoulders draped over the metal barrier. No more than five metres from the old banger. He raised one arm to wave at Billy and get her attention.

Fuck's sake.

One minute to go.

Billy started running towards the abandoned vehicle.

ZARA

Vauxhall Astra (1997 model)

The phone won't connect. But they already know where the car is. The plan is shot, but that's on the higher-ups. I had to make a call. Damage limitation. They'll never trace the car. But they could trace me, so—

I had to go.

Two more minutes, according to the watch.

I walked two and a half miles of motorway in twenty-nine minutes. Even in this heat. Drivers eyeballing me all the way along the carriageway. Up the gridlocked ramp. Same story for the next two miles. Progress has been satisfactory – until now.

Apocalyptic scenes on city roads. People eat food off the bonnet of their cars. Blankets slung over road signs for shade. Screaming babies. More abandoned cars. People can walk to find services. It's cooler, a cross wind, not hemmed in by those high barriers. Atmosphere is much the same though. Vacuum of tension.

One more minute.

Keep moving. Calm but purposeful. Swift not running. Don't draw attention. Roads are narrower here, spaces between cars plugged by bodies. Primal atmosphere. Elemental emotions. Anger. Shock. Tears. Abrupt flares of rage. Racial, sexual or just plain nasty outbursts. I walk on by. People erupt like magma finding its way up through the tarmac, through society's weak spots. It's always the same, fear brings out baser instincts. Don't slow down. Don't excuse myself. People glance up, step aside, look away. No one registers anything beyond the headscarf. There's nothing memorable about me.

Watch shows it's time—

Behind, in the distance, a boom. Now I stop. Like everyone else, I face west.

A radio plays the jingle of the six o'clock news.

Can't see anything, not even smoke rising. Must be a couple of miles away. The sound is audible because the usual background hum of traffic is gone. Then it's lost to cries. Groans. A woman pulls a child to her chest.

I move on.

THE SECOND HOUR

Chapter Twelve

6.03 p.m.

Billy hauled herself up from where she'd fallen over her own feet and crash-landed. She brushed off grit that left behind sharp triangular indents in the heels of her hands. Tutted at a rip in the knee of her jeans.

Her head snapped up in response to outraged voices on the other side of the carriageway. A knot of people around one car, listening to the radio, broke apart into groans and shouts that ricocheted up the motorway, echoed by other groups who were also tuned in to the news. A car bomb in the west of the city. Exactly as Dom predicted.

Billy eyeballed the abandoned vehicle which was only a car-length away. Its headlights the shape of doleful eyes. Like a dog. She wasn't sure if she could trust it or not. But it wasn't the six o'clock bomb.

She listened to a radio that someone had turned up to full volume. The latest blast had gone off on the hour after a vehicle pulled into the taxi rank outside a BBC radio studio and detonated. The journalists inside the building must have simply carried on doing what journalists do because they flashed the news of their own explosion within minutes.

At least hardly anyone on the motorway had noticed her fall. They were too busy reacting to the news in their own unfiltered way. Some stood with hands interlaced on the back of their heads, as though they were being arrested in a cop movie. Some were limp. Some cried. A few preferred to be alone in their cars, but most gathered for company. Others seemed completely unable to process what was happening and carried on regardless, including the lad who had made Billy do her mad dash across the carriageway to 'save' him.

He still had his arms flopped over the metal barrier, patiently waiting, and he was wearing a beanie in this heat.

'So, anyway,' he said, 'I want to report a crime.' He adjusted the hat to reveal an inch more forehead. The beanie had an infinity symbol embroidered on the front. 'I heard you're a copper. There's been a theft over here. People are getting a bit out of order.'

Her head was still ringing from the fall.

Instead of *you absolute dickhead*, she said, 'What theft?'

'Someone stole my picnic hamper. I left it over there and went back to the car to get a blanket.' He pointed behind him to the grass verge on his side of the motorway. 'And some bastard's taken it. I mean, I know everyone's short of supplies . . .'

Billy shifted her stance. She now actively needed a wee, which didn't help her patience.

'They can't have gone far,' she said.

'It's still a theft.' He gave an elaborate shrug. 'Police don't have time for anything these days.'

'There's been a fatality. It might have to take priority over a picnic.'

'I'll deal with it myself then.' He spun on his heel and marched off.

'Sir?' When he didn't reply, she let him go. He could find his own bloody picnic.

Billy checked her watch. The passing of the six o'clock dead-line made it seem unlikely that this abandoned car was suspicious after all – the modus operandi of the terrorists seemed to be driving up to a site of significant value and detonating a device, not leaving a vehicle in the middle of nowhere and risk it going off at the wrong time. Billy knew enough about terrorism to see that an attack on this scale was meticulously planned. These lot were all about the timing, the spectacle, the media coverage. They wanted the world to see, not a handful of nobodies on the motorway. If she cordoned off the abandoned vehicle so that no one started poking about with it, she had half an hour before she even needed to think about it again. Time to focus on Dead Driver. She jogged back down the road towards her crime scene.

So Chad McClusky died just before 5 p.m. Metal thing, black and the size of a skewer, in the back of his neck. No other visible injuries. Heavy stop-start traffic. No CCTV. Vehicle unlocked. American national. Two forms of identification in the name of Chad McClusky. Wedding ring that looked brand new. Arrived from Washington, heading to Howle Green, where there is a US airbase. Possible that he was connected to that airbase for obvious reasons . . .

As Billy considered that fact, she let her peripheral vision take in the Hybrid Welfare Unit, the weird CIA-style vehicle on the other carriageway. Still there, of course. Driver still at the wheel.

The van had windows along its side, but they were tinted so dark she couldn't make out what was in there. Although—

She looked away. Was that movement? A glimpse of a shadowy shape in the darkened windows. Someone in the compartment of that van?

Could be surveillance.

Could be someone watching.

But if the weird van was involved in Chad McClusky getting bumped off, what were the chances of it stopping at the exact right location on the other side of the motorway going in the opposite direction? If it had been tailing his sedan, she could understand it, but from the other side? Too much of a coincidence, surely, that the traffic would stop in just the right place?

No, she was getting spooked, that was all.

She looked directly at the black van and saw . . .

Nothing. No more movement. Just a driver wearing AirPods and tapping a fingertip on the wheel to the beat. He probably had the aircon switched on, keeping him nice and cool.

Enough with the flights of fancy.

Luckily, Pat was ready to bring her down to earth. The woman advanced between the cars to join Billy. 'These phone lines are jammed, I reckon. They do that because of the terrorists. Stop the fuckers communicating.' When Billy didn't break stride, Pat spun on her heel to follow her back to the grid. 'I was in India for the Mumbai attacks. One terrorist was giving instructions to the others by phone. Don't miss a trick, do they? What year was that? Late noughties . . . They probably use WhatsApp now.' As they reached the rope around the black sedan and stopped, the nurse hoisted her bosom and gave it a moment to settle.

'Mumbai was 2008,' said Daisy in a voice loud enough to detonate a car bomb all by itself. '9/11 was 2001. Madrid, 2004. London, 2005. Mumbai, 2008. I'm reading Criminology and Politics.' She shrugged an apology.

That explained all the questions.

'She's going to be prime minister,' said Olly.

Oh, get a room.

Billy wanted to ask Pat what she'd been doing in India. The nurse was a curious woman. But now was not the time.

'Yeah, maybe they've jammed the lines,' Billy mused. 'Or maybe the network is overloaded.'

'They're jammed,' said Foul Mouth. 'I remember the tone. I was training field nurses and we wanted to get them out on the streets to help people. But I had to call London and get them to phone my colleague in the hospital just down the road and then call me back. International lines worked, see, but the local network got jammed. Hard workers, those nurses, never bloody stopped. Christ almighty!' She shook her head at the carriageway as though it was teeming with idle English nurses.

'So anyway,' Billy said. 'I need another cordon for that abandoned vehicle.' She turned to Olly. 'Could I trouble you for more rope?'

She tailed him to his camper van. He slid open the side door, but held out one hand to keep her back.

'If you don't mind, I'm just—' Olly didn't say what he just was. Billy held up both hands. She hadn't been about to force her way into his vehicle. But now she was more curious than she'd been about the camper van. She kept her distance while he squeezed past an expensive-looking mountain bike stored in the living area.

'Where you off to, then?' Billy asked.

'Just on my way home.' His voice was crushed in his chest as he crouched to rummage inside a plastic storage box. 'Heading to my folks' manor to put this lot in storage. Then back to uni.'

With his accent, it was hard to tell if his 'folks' manor' was a Guy Ritchie-style cockney affectation or literally a massive pile in the country.

'Anywhere nice?'

'They live not far from Avon Spa?'

She'd meant uni, but whatever. Avon Spa sounded about right.

'Isn't that south of here?' Billy asked. 'And we're heading north?'

'You know your geography.'

She agreed.

'I'm taking a bit of a detour to pick up some bike parts.' Olly squeezed past the bike again and hopped down onto the road. He slid the door closed before handing her another fluorescent rope.

'Hey!'

The shout came from behind. It was followed by the incongruous but unmistakable clattering of hooves. A pony came trotting down the hard shoulder, a child bouncing on its back, no saddle or bridle or anything, just hanging onto the animal's lead rein.

'No, no, no!' Billy ran to stand in its path, waving her brightly coloured rope.

The pony veered onto the grass and broke into a canter.

'Stop!' Billy stepped in front of it again. 'Police!'

The girl wasn't going to stop, her toes turned out as she dug her heels into the animal's side, urging it on. Billy considered grabbing its rein, but there wasn't space – the pony might spin round and crash into the mother and daughter's minivan, or the kid might come off the side. She didn't want a broken collarbone to deal with too. She waved her arms and, when it became clear the pony wasn't stopping, dodged out of its way. The animal barrelled past.

'Little bugger!' Billy shouted after the pony's arse. The kid waved her hand in the air as if to signal an apology.

'It's like *Mad Max* round here.' Pat appeared at Billy's elbow.

'Kid won't get far,' she said. 'Just hope she doesn't get hurt.'

'She'll get what for on her way back, judging by your face.' Pat hooted and tugged at the thin fabric of her kaftan-shirt. 'Don't know how she's got the energy, it's too hot for all that.' She raised both arms to let the air circulate and her sleeves slipped down to her elbows. Billy noticed a tattoo under her wrist. Both wrists. Swallows.

Jailbirds.

Shit, shit, shit.

Swallows signalled women who'd served time. Pat was an ex-con. Billy had let this woman insert herself into the black sedan when they examined Dead Driver. Bloody clever. Everyone had witnessed her contaminate the crime scene.

Chapter Thirteen

6.08 p.m.

Pat Mackey flashed a grimace that revealed a row of bottom teeth like fudge. Billy's teeth had fallen into line at an early age (rather like herself), which was lucky for someone born in the seventies, when cosmetic blessings came from nature or not at all. As Pat continued to mutter foully about the phone signal, Billy noticed how she spoke in a tight-lipped way to conceal the bad teeth. And this made her seem angry. Maybe it had set the tone of her life. Billy had seen lives derailed by lesser matters.

Focus, Belinda.

So Pat was an ex-con.

Prison then nurse? Must be the other way around. Nurse then prison.

A nurse would have the know-how to administer a fatal injury. For example, by inserting a skewer in the back of a man's neck.

Billy believed in giving people a second chance. But, still, she would have to ask.

'How long did you serve?'

'Fuck's sake.' The nurse sucked her lips inside her mouth and released them with a smack. 'A twelve year. Got out last summer.'

Twelve years.

Must have been serious.

Here she was, stuck in a traffic jam with an ex-con and a murdered man. Time seemed to have stopped. The unusual situation had given her a weightless sense of immunity. It was like the mood she'd succumbed to on the long flight home; both vulnerable and protected. She'd eaten and drunk everything the stewardess presented according to some unfathomable airborne timetable, as passive as a child or a patient. Or a prisoner. Ice cream at breakfast. A pointless packet of peanuts. A shiny green apple in a napkin. This untethered mood had stayed with her, along with the thin coffee in her bladder. But now she had to snap out of it.

'Why the tattoo?' Billy was distracted by a blur in the sky – *helicopter?* – but it was a distant buzzard, patrolling in silence. A sign of life from beyond the perimeter of the motorway.

'Rite of passage, I suppose.'

'I'd have thought you'd want to forget all about prison,' Billy said.

'Some things won't be forgotten.' Pat spoke in the most placid tone she'd used so far.

A waft of hot rubber reminded Billy of the scooter incident and the truth in those words.

'You okay, love?' Pat asked. 'Look a bit peaky.'

'Jet lag,' said Billy.

Pat bobbed her mouth as though that might explain it, at a stretch.

'Do you know this man?' Billy asked. 'The dead driver?'

'Never seen him before, alive or dead.'

'Do you know the name Chad McClusky?'

'Weird sort of name.'

'But do you know it?'

'Never heard that name in my life. Not until you said it on the phone to your mate, the other copper.'

99

Canny old bird, didn't miss much.

'What did you see when the traffic first stopped?' asked Billy. 'Before we discovered the body?'

Pat frowned at Dead Driver. 'Not much. News came on the radio about the terrorists. Traffic stopped. More news. I got annoyed. Heard the screaming.'

'Did you see a motorbike? Would have been right after we stopped.'

Pat frowned so hard her lumpen teeth appeared. 'Nup.'

'Sure?'

'Sure.'

'Because I remember a bike revving,' Billy offered. 'I had to turn up the radio to hear over the racket.'

'Maybe, I . . .' Pat scrubbed at her nose with the back of a hand. 'I was pretty irritated. By the hold-up.'

Irritated. The art of understatement. The woman had been spitting out words like flies had flown into her mouth.

'I promised I'd be somewhere and I'm not, am I? I'm stuck here. So I got annoyed. I may have been distracted. And a noisy bike on a motorway isn't worth noticing, is it?'

Everyone had been distracted. Including Billy, with her hot sweats brought on by driving. It seemed crazy now she was stationary. A fear that fired up with the car engine. Everyone had been caught up in their own business, oblivious to what was going on around them. Other people little more than objects to be navigated, like traffic cones, entire lives reduced to collisions that had to be avoided.

'And where were you off to?' asked Billy.

'I volunteer. For the homeless.'

'A refuge?'

'Drop-in centre. Helping them get access to services, doctor's appointments and benefits and that. Stuff they have to do online and don't have a clue. Half of them are illiterate. Some of them have been inside too, never had to deal with computers or paperwork. Someone did it for me when I came out, got me straight, now I'm passing it on.'

'That's very commendable.'

'Except I'm not there, am I? I'm stuck here.' Pat crossed her arms and scratched the underside of her wrists. The swallows.

'I'm sure they'll understand in the circumstances,' Billy said.

'They won't. They'll think it's just me being me.'

'I'll vouch for you,' said Billy. 'If needs be.'

Pat stopped scratching and turned to glare backwards down the carriageway.

'There was one thing,' she said. 'That lad in the Range Rover.' She pointed back past the black sedan.

'The estate agent?'

'Yeah, I'm not usually one to snitch but he picked something up off the hard shoulder.'

'Did you see what he picked up?'

'It looked like a phone.'

Pat made to accompany Billy as she moved off towards the estate agent.

'I'll be fine on my own,' said Billy.

'Call if you need backup,' said Pat. 'Woman on her own, shouldn't have to deal with this shit, copper or no copper.'

'I'll be fine,' said Billy.

'And I know when I'm being told to fuck off.' Pat trailed back to her own vehicle.

The nurse already knew more than she should. Billy watched her depart and then approached the Range Rover. Only a car-length behind the black sedan, the estate agent had the best possible access to the dead man.

Pat aka Foul Mouth

VW Golf (2010 model)

The sun visor has a mirror. And a light. Natty. I'd forgotten about the light, quite the little luxury back in the day. But back then I wasn't confronted by a face the colour and texture of a digestive biscuit. Twelve years, gone. Nothing to show for it except sharp little lines stabbing my top lip. Hardly looked at myself in all that time. Used to, though. Had time to waste. So I thought. Loved to put on my uniform and my trustworthy face. Same face, different woman. A lit-up mirror doesn't seem so natty these days.

Could sell the car. Bank the money. But that seems contrary after storing it all these years. Only 6,000 miles on the clock, how long did I even get to use it, before I went down? Bought it in the summer to drive Janey to France, that's when I found out what she'd gone and done, and she was dead by the autumn—

Stop grinding your teeth. Squeeze the wheel instead. Tight and tighter and release. Better than grinding, don't get free dental on the outside.

Weather had been like this when it happened. Unseasonably warm. Hazy days. Hazy daze. Can't remember the actual deed. You'd think you couldn't forget such a thing. But anger is a veil, isn't it, and I was trying not to let Janey see through. Only made it worse, holding it inside, the fury, the hurt, compressed into a white-hot spike – I can remember that feeling if nothing else – hard as metal, a needle point, consuming, distracting, and that's when it happened. She was asking for it. So I did it. And there it is.

Grinding again. Do the thing with your hands. Palms up. Make a fist. Clench. And. Release. Let your fingers relax—

Can't remember the deed. Only the park after. Needing fresh air but getting the reek of rotten leaves. Dumping the sharp in a bin. Police never did find it. Not that it made any difference, it was pretty bloody obvious what had happened, they didn't need much hard evidence.

Fist. Clench. Release.

I can see the dead man in the wing mirror, two cars back. Knew he was dead the minute I clapped eyes on him. Know a dead face when I see one. Seen enough in my time. It's hard to explain, how they look vacated, all the fight gone out of them, you just know when you know.

It's not peace. It's surrender.

Janey surrendered in the end.

In a way, we both died, but I'd do it again. Helping people is my calling. Speaking of which, the policewoman is with that fella with the beard. Looks like he's drawn it on with a Sharpie. What do they think when they look in the mirror, these people? Same with the Botox lips, how did that come to be normal? A lot changes in twelve years. You expect it of clothes, but not people's actual faces. And don't get me started on eyelashes . . .

What's she up to with him? She might need help. I'm ready, just in case.

Chapter Fourteen

6.13 p.m.

Billy texted Dom as she crossed the carriageway, asking him to look up Pat Mackey's criminal record. Again, WhatsApp displayed a tiny clock symbol, the message wasn't going through. The previous text was still waiting too. She'd have to find some other way. But first, the estate agent. The waft of citrus cologne from his Range Rover carried a furtive note of root vegetable that Billy didn't want to ponder. She stepped back half a pace. Behind her, on the other side, a trio of boys were heading a saggy football from one to the other with a series of exhausted flumps. It made her think of medieval villagers amusing themselves with a pig bladder.

Don't think about bladders.

She turned back to the Range Rover.

'Did you get out to retrieve anything from the hard shoulder?' Needing a wee brought on an urgency that she found hard to curb.

'Earlier?' Carl said.

Obviously.

'Yeah, I had a bit of a' – he shrugged – 'a blip.'

'A blip? Do you mean when you got out to retrieve something from the hard shoulder, or when you failed to tell me when I questioned you?'

'I didn't realise I was being questioned. You're not in uniform.'

'What do you think I am, a traffic warden?' He'd had two chances to tell her about whatever he picked up off the hard shoulder. And he hadn't used either of his chances.

'Sorry, I . . .' He stroked his facial hair. It was trimmed in the shape of an anchor under his lips, but so finicky and fussy and precise that it looked like a child's cut-out disguise. 'It was nothing, I forgot to mention it.'

'What kind of nothing?'

'Just something I threw out and then regretted, so I picked it up again.'

'Like what?' she asked.

'What?' he said.

'It's a simple question.' The pressure on her bladder made her desperate for any kind of release. A rabbit punch might wake this guy up. Win-win. Instead, she slid her thumbs under the waistband of her jeans to make more space and said, 'What did you throw out and pick up again?'

'A wrapper. Plastic. I don't like to litter.'

'What kind of wrapper?'

'Chocolate.'

'What chocolate?'

'Snickers.'

'Where is it?'

'What?'

Christ. 'The Snickers wrapper – where is it now?'

'I . . .' The facial hair received another petting. 'I dunno.'

'Presumably, you didn't throw it out, pick it up, then throw it out again. So where is it?'

'I dunno.'

'It wasn't a chocolate wrapper, was it?'

'What right have you got—?' he started.

Billy planted her elbows on his window, pushed her face right into his sex-pest goatee, breathed his fetid air and asked, 'What was it that you picked up from the hard shoulder, *bab*? If you like, I can haul you down the nick as soon as we get moving, and you can tell me there.'

He pointedly pressed his head back into the headrest. 'It was a phone, okay?'

Billy stood upright. Pat had been correct. Didn't miss much that one.

'Go on,' Billy said.

'It was just this little Nokia. We don't give our personal numbers to clients, security measure, so they give us these little burner phones. Makes us look like drug dealers, but the agency is cheapskate, what can you do?'

'And why was it on the hard shoulder?'

'Like I said, I threw it out, then picked it up again.'

'Why throw it out?'

'Because I was annoyed. I was on my way to a big commission. Second viewing, thought I'd get an offer, maybe even close it today. But I got stuck here, and while I was trying to rearrange the viewing, I got a message to say the seller had got a cob on and accepted an offer from another agent. So I'd just lost ten grand, my window was wide open, and I threw the phone. Then I thought that was stupid, so I got my tail between my legs and picked it up again. Luckily, it wasn't broken. End of.'

'You had your window open on the motorway, with all the fumes?'

'Yeah, bit of breeze. It's hot.'

'Where did this fit in to the timeline of seeing the motorbike?'

He thought about it. 'The motorbike was there right at the start and raced off a minute or so later.'

'So you would have got out of the car how long after we stopped?'

'A few minutes?'

Billy squinted up at the sky. The sun was so white it stung. She blinked away black spots. 'If it had only been a few minutes, how come the house was already sold?'

He scrunched up his eyes and dragged his fingernails through his beard as though the question physically hurt. 'I told you. Owner lost his rag, thought I weren't coming, and sold to someone who'd viewed earlier. It's a seller's market, too—'

'Can I see this phone?'

He shrugged his hands and reached for a Nokia from the centre console. Billy made him stop and took photos of it in situ, zooming in on a scrape across the buttons which would be consistent with it being tossed onto the road. Ideally, she needed a Faraday bag to prevent electronic interference. Could you remotely wipe a burner phone? Maybe that was only smartphones? She wasn't sure, normally she'd take no chances, but today was not normal. Instead, she got out a poo bag, licked her fingertips to rub it open, and held the rubbery plastic outstretched. He gave her stinkeye but dropped the burner in and she tied it up. She wrote all the usual details and an evidence number on a scrap of paper and double bagged the item.

'I'm going to have to keep this phone.'

'What?'

'Murder investigation. What's the number?'

He heaved a huge sigh that blew out his lips, but found the contact on his smartphone and read aloud the number. Billy noted it on her pad.

'PIN number?'

He shook his head. 'Everyone's gone yampy,' he muttered, but told her the PIN.

Right on cue, the sandy-coloured dog came bowling past and Billy didn't react quickly enough to grab its lead. It bounded up the carriageway, tail and tongue lolling from side to side.

Billy stamped back to her vehicle and shut the burner phone into her glove compartment. In her peripheral vision she eyeballed Carl, but he was sitting peacefully in his gaudy car. He adjusted his pale blue tie and stretched his neck side to side, as though he'd just triumphed in a skirmish—

'Sergeant Kidd!' Daisy was flanked by Olly and Charlotte. 'I heard a scream.' She pointed at the CIA van on the far carriageway. 'It came from over there.'

Carl aka Brummie Sex Pest

Range Rover Evoque

Can't stop touching this beard. Feels like another person. Not that I've ever touched a man's beard, no offence, like, just not my preference. And the hairy parts of a woman don't feel like this, except for that one lady with long hair that had a shaved bit underneath. She felt similar, a surprise when my hand went round her neck, erotic in its own way. But beard hairs are thick and blunt. Like blokes, the wife would say.

What the hell will she do when she finds out about this?

Can't hardly think.

My head's like . . . full of brambles. Making my eyes prickle. Years since I cried. Christ on a bike, my stomach is churning too. And there's a rough patch under my chin that hurts when I drag the hairs against the grain. Blind spot. Or cancer. Runs in the family. Should get it checked. Can you get check-ups in prison? Won't come to that.

Will it?

Should have got rid of the beard. But shaving in a motorway service station isn't exactly the way to keep a low profile and how would I have explained to the wife if I'd turned up clean-shaven? Oh, yeah, no, you're alright, bab, I've only spent a month growing a beard even when you said it made me look like a paedophile and I've been pruning it into this fucking stupid style but now I've had a change of heart and shaved it off in the middle of the day in a public lavvy. Hardly going to believe that, is she? She'd think I was having an affair.

Ironic.

That's a song.

She got stuck in a traffic jam when she was already late.

That's not ironic. Just bad planning.

At least I got rid of the baseball cap, so that can't be matched to any CCTV. There must be surveillance cameras. Stomach's churning. If there's CCTV they'd have seen me dump the cap. They could be fishing it out of the bushes right now. Gloved hands. Evidence bags. Like this copper. I should have got rid of the beard. Christ alive, my guts are hotter than the bloody asphalt.

Chapter Fifteen

6.18 p.m.

It had been a short scream. 'Like someone missing a step on the stairs but catching themselves before they fall?' Only Daisy had heard the scream because she'd gone back to her car to get a drink from Olly's water bottle stowed in the door. She showed Billy the bottle as though that proved her story.

Billy could hardly focus because her mind was having a Pavlovian reaction to the slopping sound of the water, and all she could think about was her overburdened bladder. She found it hard to concentrate when she needed a wee. Now she'd have to haul herself over the central reservation. Like a barrel going over Niagara Falls.

Stop thinking about water.

The metal barrier separating the two carriageways was chest height. Billy got Pat to give her a leg-up, straddled the rail and flopped over to the other side. Her jeans and belly were left smeared with road grease. Having surmounted the obstacle, she felt more confident than ever that whoever killed Chad McClusky didn't escape over the central reservation – not without being seen. It was not a subtle manoeuvre.

The driver inside the black Hybrid Welfare Unit van didn't react to Billy appearing over the central reservation right in front of him. He just stared straight ahead.

If you drew a face very neatly onto a football, it might resemble this man. Skin so smooth it looked tight. He was of an indeterminate age – if pushed, Billy would estimate early thirties – dark hair closely cropped, but not so much that he could be described as a skinhead. Average features that didn't come together in a way that was either ugly or good-looking. Or memorable. Handy for a spook. Ideal, in fact. She'd always wondered how someone who looked like James Bond was supposed to go unnoticed. Bit of a plot hole, that.

Focus, Billy.

Van Man wore a plain black T-shirt. Billy couldn't see what else because he remained sitting up high in his van. He let down the window when she gestured at him.

'Hello,' he said in an unplaceable English accent. He didn't ask questions or offer any comment about the dramatic events over which he'd had a perfect view for the past hour and a bit.

'What's your name?' Billy asked.

'Bob.'

'Bob? Really?'

He shrugged.

'And your surname?'

'Who's asking?'

'I'm a police officer.'

'Like I said, who's asking?'

Like that, is it?

'Was there a scream?'

'When?'

'A few moments ago. Someone reported a scream.'

The driver pulled his mouth down and shook his head. 'I was listening to my phone.'

'What you listening to?'

'Podcast.'

'Which one?'

'True crime.'

'Oh, I like those. Which one?'

'Someone got convicted of a murder they didn't do.'

'That's all of them.'

The driver smiled begrudgingly. He leant forward to tap his phone. He held it up and Billy saw the familiar red-and-black logo of a show that had kept her company during frequent beachside walks in Perth.

'No spoilers.' He put his phone back into its holder, taking the time to plug in the charging cable.

'So, this scream?' asked Billy.

'Don't think so,' he said.

'Don't think so what?'

'I don't think there was a scream.'

Billy nodded. 'What's a Hybrid Welfare Unit?'

'I'm just delivering the van. Hybrid means the fuel. It's part electric.'

'And welfare?'

'Dunno.'

'Can I take a look in the back?'

'Better not,' he said.

'Why?' Billy was all surprise.

'Not my van, is it?' the man said.

'Does it matter if I take a look then?'

His chest rose and fell in a tiny sigh, then he flung open his door – forcing Billy to take a hurried step back – and got out. On the road, he was trim, his T-shirt designed to show off his chest. 'Is this a stop and search? You got to tell me what it's about. What you're looking for.' He stood slightly too close. His features were relaxed, even slightly amused. 'And can I see your warrant card? I have the right to see your identification.'

Shit.

Billy said, 'I'm off duty.'

'You've been retired.'

How did he know that?

'First thing you said when you got out the car.' He gave her a quick up and down, eyelids flicking. 'You don't look old enough to be retired.'

Flirting? Challenging? He was hard to read.

'Or was it the misconduct that got you chucked out?' he said.

Challenging, then. 'You've listened to too many podcasts, Bob.'

'Don't suppose you can go back into the police after you ran that scooter off the road?'

She hesitated a moment too long.

He smiled broadly, knew he'd caught her off guard.

What the actual, does he know . . . ?

Billy shook her head; once, hard. 'Can I look inside your van or not?'

Now he shook his head, but slowly, sadly, as though it pained him. 'Not without a warrant card. It's not my van and it's an expensive bit of kit. If anything happened to the equipment back there . . . I'm risk averse, see?' He raised his palms to God as his witness. 'In some ways, it's a blessing. I'm a safe pair of hands. But not everyone can say that, can they, officer?'

His relaxed expression was back. Talking to this man was like talking to a vicar, there seemed to be an undercurrent to everything he said.

Billy had a strong urge to get back to her side of the carriageway. All of a sudden, it seemed like home ground.

'If I hear any screaming, I'll call you. Sergeant Belinda "Billy" Kidd.'

She frowned at him.

'You were in the news,' he said.

Her heart flopped over like a badly tossed pancake. Now she could classify his expression. It was pure menace.

'You got no right to question me.' He lowered his voice. 'This is a murder case.'

'You've got a nerve, running around on the road, pretending to be in charge, after what you did to that kiddie.'

'It wasn't a kiddie.'

'Teenager. I googled you. And I read all about—'

'That's enough,' said Billy.

He laughed. 'Thought it might be.'

Billy heard his van door slam as she clambered back over the central reservation.

Bob aka Van Man

Mercedes Sprinter (Custom Fitted)

Can't hardly concentrate on this podcast anymore. And it's obvious who done it. The very last person you'd expect, that's always who done it. Some nobody in the background of episode one whose presence you barely noticed, like an old busybody or a delivery guy, but it is *never* the person they make you suspect in episode three of a ten-part series—

A knock in the back of the van. I feel it as much as hear it. Thumb down the volume.

Nothing.

After a few seconds, I realise I've tensed right up. No need for drama, it's all in hand, I make my muscles soften.

Turn up the volume again.

True crime gets samey. I've overdone it. I'm like that, obsessive, can't pace myself, bingeing a whole series in one go during long stretches on the road, redeploying 'cargo' late at night when the music on the radio gets weird and electronic, turning to real human voices when I'm stuck in the cab for hours on end. Should be able to write one myself by now. I'd have the material too, stuff I've seen. Maybe I will. When I get out of this game.

Crack the window, better to hear them on the other side of the road. The voices carry. She got herself back over the central reservation. Looked like a hog on a spit, but she managed it, give the woman her due. Bossy sow. The others hang around her like piglets. If push comes to shove, I might have to shatter their illusions that she's a mother hen.

See, I could do a podcast if I wanted.

From the back of the van, another knock.

Raise the fist. Wait—

Scratching, a whine.

I pound low on the metal panel that separates the cab from the back of the van.

Bam, bam, bam.

They need to keep it down. It's like they don't understand what's at stake here. The noise stops but I pound once more to make the point.

Bam.

Chapter Sixteen

6.28 p.m.

'Tell me if you see any movement inside that van.' Billy spoke in a low tone to Daisy and Olly, whose position on the outside lane afforded them a view of the other side. The black van in particular.

'It's like something out of the FBI,' he whispered.

'CIA,' Daisy corrected him. 'Security sources suggest that half of all CIA surveillance designed to prevent terrorist attacks on American soil is carried out here in the UK.' She waited a beat for them to catch up. When they didn't, she added, 'Because we breed so many terrorists.'

'Right,' said Olly. 'God, that's so interesting.'

He must really fancy the girl.

Although it did make Billy wonder if there could be some connection between this murder and the terrorist attack. She let the idea pinball around her mind for a few moments, but it failed to strike any jackpots. Apart from the obvious fact that they were only stuck in this gridlock because of the terrorists.

Billy checked her watch.

She grabbed Olly's rope and dashed off down the motorway. Her legs fizzled like human Alka-Seltzer. She reached the abandoned car and worked quickly to ensure the surrounding vehicles were evacuated, that no foolhardy passenger had sneaked back from a safe distance. But there didn't seem to be as many people on the grass verge as before. She started laying the rope across car bonnets to mark the cordon.

'What's happening?' yelled Gammon. His face was sunburned to the colour of a boil.

'Where is everyone?' she yelled back.

'Sitting in the horsebox, in the shade.'

'Good.'

In a small hatchback, about four rows back from the suspicious car, she found an older woman still at the wheel. The driver opened her window an inch. Her eyes were two swirly marbles. But her bright white hair was cut in a trendy bob.

'I need you to vacate—' said Billy.

'Are you the WPC?'

As Billy drew closer, she saw that the woman had stripped to the waist and was wearing only a bra. Billy kept her gaze well above the collarbone.

'We haven't used that title this century,' said Billy. 'I'm Sergeant Kidd.'

'Stay back, then, stay back!' The woman scrambled in her seat as though she could push herself backwards through it. Billy held up her palms. Assuming the woman wanted privacy in her state of undress, she stopped a few metres shy of the window.

'Don't touch me!' The woman was working herself up. 'You'll have touched whatever killed that fella, got it all over your hands.' The woman's window closed up until there was barely a centimetre gap.

Billy glanced at her hands. 'I don't think there's any risk—'

'If you don't know what killed him, you don't know what the risk is. I've been sitting here, thinking about it. I reckon it's Novichok.'

'The nerve agent?' said Billy.

'You should self-isolate.'

'I need you to leave your car, madam, and move further away from this suspicious vehicle.'

'I'm going nowhere. We don't know who's out there, roaming about. My doors are locked. You can't shift me, there's no point trying.'

'There's a suspicious vehicle.'

'Twaddle. If it's not gone off, it's not going off. It's him we've got to worry about.' The woman jabbed a knotted finger in the direction of the black sedan. 'I'm staying inside with the air conditioning switched off so the toxins can't get inside.'

Billy checked her watch. 6.29 p.m.

'It wasn't Novichok that killed him.'

'You want to wash your hands. And you want to warn those poor people.' She pointed in the vague direction of the cars around the sedan. 'They're sitting ducks.'

If it was bloody Novichok I'd be frothing at the mouth.

And so would all the drivers in the grid, Charlotte, Pat, Daisy and Olly . . .

Poisoning was one thing she was not worried about.

She ran the rope behind the woman's car and over to the hard shoulder.

Nothing she could do about her if she refused to move.

A brown blob in the distance revealed itself to be the pony barrelling back this way.

'Make sure people stay out of the cordon, okay?' Billy shouted across to Gammon.

He only grunted in return.

Billy set off along the hard shoulder in the same direction that the traffic should be flowing. She got a good six vehicles past the suspicious car and waited on the red tarmac for the pony to arrive.

6.30 p.m.

She glanced back at the abandoned vehicle. Its headlamps doleful. As though waiting for the inevitable.

6.31 p.m.

Nothing.

Good.

So it was just an abandoned car. False alarm.

A human bark snatched her attention to the young men playing with the deflated soccer ball on the other side. Their grunts and lunges struck her as more unruly now. The game had developed the rhythm of a tribal dance.

All humans live in tribes.

A guide had said that to her at a museum in Australia, triggering an onset of loneliness so deep it felt like dehydration.

What about a person who lives alone, no family, no tribe? Are they less human?

The high jinks of the men with the football were tribal. She could imagine them going on the rampage. They would be the first to cross the barrier between carriageways, should chaos descend. The first to expand their territory by overwhelming the weak.

She must not let chaos descend.

The tribal lads gave a war cry. If Van Man blurted out what he knew, she would lose her little territory, her little authority. She would lose control of this crime scene. But that wasn't all that rattled her. The fact that he'd uncovered her past made her realise that she hadn't left the misdemeanour behind. It accompanied her like a limp. Eventually, it would exhaust her.

Suddenly, she very much wanted to hear a friendly voice. Almost as much as she wanted a pee, and a glass of cold water, in that order. To her right, in the middle lane, was a taxi.

A taxi must have a radio.

The pony was still a way off, moving more slowly now.

The passenger window of the taxi slid down as she approached.

'Alright, love.' The driver offered her a mint humbug from a jar and after a moment's hesitation, she took one. The taxi registration displayed in the windscreen identified him as Mandeep Singh, a photo confirming his identity. Billy introduced herself. A pack of playing cards was arranged on the passenger seat. Mandeep was playing poker.

Billy pointed at them. 'Three of a kind.'

He put a finger to his lips and nodded backwards to indicate the rear seats:

'Don't tell Jack, he keeps on beating me.'

Billy peered into the back of the taxi to see a young black boy, perhaps ten years old, lying full length on the seat. He had a hoodie pulled tight around his face and a handful of playing cards in his grasp.

The window slid down for her to speak to him, and Billy nodded her thanks to the driver.

'Hello, Jack, I'm a police officer. Aren't you hot with that hoodie done up?'

The boy nodded but didn't raise his eyes off his cards.

Mandeep said, 'His mum got stuck at work and sent me to collect him from after-school club. The posh school out past the airport. First time he's been in a taxi and then we go and get stuck in traffic. He's alright though. I've been keeping him busy.

And I put the aircon on every now and then. Can't work out how long the engine will run before I'm out of electric.'

'You'll be alright for a while.'

Billy hunkered down to talk to the boy. He pushed himself up to a sitting position and showed Billy his five cards. Each one was a heart. He whispered: 'We're betting for humbugs.' He opened his pocket to show Billy about a dozen sweets.

'Do you need any help, Jack?' Billy asked. 'Are you worried?'

'No, I can beat him,' he whispered. 'I go to poker club. And chess club.'

'Smart kid. Shout for me if you need anything, my name's Billy.' He nodded and she went back to the passenger window. 'Can you speak to your control and ask if there's any way to contact the police station? The phones are still down and I need to speak to my colleague.'

'Easiest way would be for her to relay a message.'

Billy gave him Dom's name and landline number, then pushed herself away from the car to go to the hard shoulder, where the pony was now approaching.

This time she spread her arms out wide and made a lunge for its rein.

The girl pulled up to a halt, the pony skittering side on, bum to the road.

'What did you find?' Billy asked.

'Am I in trouble?' The girl's defiant expression suggested she was delighted by the idea. Tendrils of fine blonde hair were stuck to the velvet of her helmet. She wasn't much older than Jack.

'I'm a police officer and I want you to go back to your horse box and stay there. It's dangerous, riding up and down.'

'I won't tell you what I saw then.' The girl held Billy's eye.

'What did you see?'

'Nothing.' A smile pinched her cheeks into rose buds.

'Nothing-nothing? Or "I'm not telling you" nothing?' said Billy.

'Nothing-nothing. Just this big fence all the way to the tunnel.'

'I could have told you that if you'd stopped when I asked you to. Did you get as far as the tunnel?'

'No, a nasty man told me to go back.'

'Sensible nasty man. Did you see any police down there?'

'Just more cars.'

The emergency services must be at the other end of the Dead-wall Tunnel. Miles away.

'So now will you listen to me and go back to your horse box? You need to let that pony cool down, he's frothing.' Sweat had turned to something resembling shaving foam under its belly. From behind, Mandeep shouted her name. Billy released the pony's rein and it swung its big rump round and trotted off.

As she approached the cab, she heard a woman's voice on the radio.

'I'm run off my feet here.' Even the static sounded cross.

Mandeep rolled his eyes at Billy. 'How can she be run off her feet when nothing's moving? But she loves a police show, won't be able to resist talking to you.' He pressed and held the button on his handset. 'The detective is here. It's about a murder. Please connect us now.'

'I'm not a detective,' Billy said. 'I'm a sergeant.'

'She loves detective shows,' he repeated.

Don't they all.

The radio gave a loud buzz. 'I got him on the line, the other detective,' the woman's voice.

He's not a bloody detective either.

The controller's voice: 'I'll hold the phone up to my microphone, don't know if you're going to hear much but go ahead, please, detective.'

'Superintendent Day?' said Billy.

'Sergeant Kidd,' said Dom. His voice was like a hand on the shoulder.

Mandeep passed her the handset and went to stretch his legs. Billy dropped into the passenger seat.

'How are you, Billy?' said Dom.

She caught a sharp breath in a hand cupped over her mouth. Didn't want him to hear the half-laugh, half-sob. She flipped her hand like she was flicking it away.

'I need the loo.' She was not desperate, not yet, but if her bladder had a warning gauge, it would have just moved to amber. Sitting down only made it worse.

'Can't you go?'

'In front of everyone?'

'Isn't there a verge?'

'It's six feet wide. Who's going to trust a police officer who tiddles in front of them?'

'Can't you ask them to turn away? Or squat between cars?'

'Have you ever had to ask suspects to turn away while you piss on the carriageway between two cars?' She remembered Jack in the back seat, but he was amusing himself with solitaire, hopefully tuned out from her inappropriate language.

'No, Billy, look—'

'It's alright, Superintendent, I don't need you to go all HR on me.'

His sigh crackled down the line.

Don't get pissy with Dom. He's the only one on your side.

In front, a metallic-blue Volvo estate. Must be thirty years old. Pristine.

'Do you remember your dad's Volvo?' Billy said.

'Worth a fortune now.'

'Surely not? I'm looking at one. It's a shoebox on wheels.'

'The 240? Absolute classic. Don't tell me you've forgotten the velour upholstery?'

She hadn't. They'd gone everywhere in that Volvo. Dom hadn't been a big drinker – booze gave him migraines, plus Billy suspected he didn't like being out of control – so he was designated driver. Always had a car full. The passenger seat of his dad's Volvo became Billy's spot. Earned through longevity or because she gave potential usurpers the evil eye. Certainly, the passenger seat gave her special status. Until she realised that Dom was always the one in the driving seat. And there was always someone else, some new girl in the back, making eyes at him in the rear-view mirror and getting secret smiles in return. Billy got Dom's public smiles, but never those secret ones.

Long before the phrase 'friend zone' was invented, she got stuck there and never plucked up the courage to find her way out. Looking back, she'd never plucked up the courage to do a lot of things. She'd spent a lifetime in the passenger seat.

Focus, Billy.

'Anything on my number plates?' she said.

'I messaged you before but I guess texts aren't coming through. An alert came up for one of the vehicles about twenty minutes ago. One of them was reported stolen this afternoon.'

Billy looked along the carriageway towards her grid. In the outside lane, Daisy and Olly stood close together. Pat was in conversation with the landlord, despite her animosity to him at the start. The mother and daughter were still shut away, likewise Charlotte in the SUV. The car with the child seats stood with its driver's door open; Billy scanned the carriageway and spotted a man peeing behind a tree on the verge. Lucky bastard.

Could any of these people have committed this crime?

It felt so brutal, so personal.

And yet these people were strangers – weren't they?

Except perhaps Daisy and Olly, there was something between those two.

She wished they would all stay in their cars.

She didn't want them . . . colluding.

'Belinda?' Dom's voice was gravelly with static. 'It's the black SUV. An Audi. The owner, man by the name of Chester McVie, says his wife stole it along with valuables from their family home. She's done a runner.'

Charlotte McVie. Aka Parking Ticket. Driver of a stolen SUV.

What kind of man calls the police on his wife? How can you steal valuables from your own home? And how stressed would it make you, if you'd just nicked a car, to then get stuck in a traffic jam next to a murder victim? Enough to get you on the Rescue Remedy.

Chapter Seventeen

6.37 p.m.

The inside of the SUV was fastidiously clean. Not like Billy's car, which smelt like old dog even though she didn't own an old dog. Billy climbed into the passenger seat when Charlotte invited her in. The cabin of the Audi was clean to its crevices. No crumbs in the creases of the seat, no crisp packets in the door, not even a tangle of chargers, just one slim cable held in a neat figure of eight by a fussy little clip in the centre console.

It must take a regular valet to keep it this clean.

Billy glanced into the back seat. In the footwell lay an out-of-place shoe. A child's football boot.

'How many kids?' Billy asked.

'Three. Little terrors.' Charlotte hiccuped slightly.

No car that carried three kids had any right to be this clean. It must take a regular valet *and* strict rules about what those kids can do. No finger-painting the windows in yoghurt, for example, like her great-niece in Australia.

'How old?' Billy asked.

'The twins are two and I have an eight-year-old.'

'Where are their child seats?'

'In my car. They don't come in here.'

Why own an SUV if you don't use it for the family?

'This vehicle belongs to your husband?'

'Yes.'

'Any idea why he reported it stolen this afternoon?'

The woman's eyes skidded all over Billy's face.

Then she started nodding and gave a breathy laugh. 'The silly bugger must have got back early from the airport and flown into a panic – he was away for the night. Wouldn't think to put on the radio and see what's happening. I guess because he can't call me . . .' She indicated her useless phone.

'Seems a bit extreme to call the police. Do you often drive his car when he's away?'

'No, but I needed to move some things and his car is bigger.'

'What things?'

Charlotte arched her back to reach into her cardigan pocket and produced the tin of Rescue Remedy sweets. She offered one to Billy, who declined, and popped a pastille into her own mouth.

'I was having a clear-out,' said Charlotte, the sweet tucked into her cheek. 'Kids' stuff. Is there any point to all these questions? You're making me feel like a criminal.'

'You're driving a car that's been reported stolen.'

'It's clearly a misunderstanding.' She shifted the sweet with her tongue to the other side.

'Where were you taking it?'

'What?'

'The stuff you cleared out.'

'Charity shop.'

'Which one?'

Charlotte sighed impatiently. 'There's a depot, where you go and deliver your stuff and they distribute it. Better than dumping it outside the shops in the high street where it gets ruined in the rain and makes a mess on the pavement.'

Billy had to concede this point.

'What's the name of the depot?' she asked.

'Ah . . . just Oxfam?'

'What's the address?'

Charlotte sighed again. 'There's a dead man. Don't you think you should be doing something about that?'

Billy let realisation dawn.

It took Charlotte a moment, but no more than two. 'You don't think . . . You wouldn't even have noticed him if I hadn't screamed.'

'I think we might have found him eventually.'

'This is ridiculous.' Her eyes puddled. 'There's a massive terrorist bomb just gone off, this *spectacular* as they're calling it, as though it's a sound and light show, and all you care about is my car.'

'What do you mean *spectacular*?'

'It's on the radio, I had to turn it off, can't listen, it's horrible, in my city.'

'I've been busy. What's happened now?'

'They're saying the car bombs were just a countdown to the big one in the centre that's just gone off.'

Billy had thought help would get here some time. But with this latest escalation—

Stay focused.

Nothing she could do about the bombing.

Charlotte went on, 'The train station is just . . . gone. Sounds like pandemonium.' The yellow pastille slipped from between her lips and she caught it on her chin with a finger and popped it back in. She wiped away a fleck of dribble and this humiliation seemed to tip her over the edge. 'And all you want to talk about is this car. Fucking police.' She shifted in her seat to face away from Billy. 'Get out. Unless you have a . . . a warrant . . . or some reason to arrest me. Get out of my car.'

'This vehicle has been reported stolen.' Billy kept her voice light. 'I only want to ask a few questions.'

'Oh, do you? Pathetic!' She spun back in her seat and this time she did spit out the pastille. It landed on her own knee and she left it there. 'Questions after questions, and nothing happens. Or, I suppose, as this is a matter of *property* now, a man's property, and insurance money and whatever, maybe you will do something. Because you police wouldn't want anyone to get away with crimes against *property*, would you? That would make your crime figures look bad. But what about people? Maybe it would be better if we went back to a time when women were the *property* of their husbands, and then we might get some protection as . . . as chattels.'

'Mrs McVie—'

'Fuck off out of my car unless you've come to help me.'

Billy stayed put. 'Do you need help, Mrs McVie?'

'Stop calling me that.'

'What shall I call you?'

'Nothing, because you're fucking off out of my car.'

Billy took a breath and released it slowly through her nose. 'There is something going on here that I don't understand.'

'Oh, you think?'

'Yes . . . Charlotte. You seem distressed.'

Her eyes flickered closed and then opened wide, as though she was being very patient with Billy. 'I told you before. I left my kids for an hour, and that was three hours ago, and there's no sign of us getting out of here. I'm sure you have kids, you know what it's like. And now this ridiculous . . . stolen . . . It's just a misunderstanding.' She flicked the tears off her face with one swipe as though she was ripping off a plaster.

Billy didn't want to get drawn into the subject of kids – her tragic story would hardly ease this mother's anxiety.

'There's a taxi driver a few cars back. He just got his controller on the radio to contact the station for me. I'm sure he'd help you to call your babysitter, if it would put your mind at rest? I could ask him for you.'

Charlotte placed both palms on her face and held them there for a long moment. Then she scraped her skin clean of tears with a finality that told Billy the meltdown was over.

'Okay.' She sniffed a couple of times and straightened her clothing. Then she got her phone and scrolled through her messages to find the number. While she did that, Billy watched the road ahead. Pat, Heathcote and the two youngsters were in a huddle. What were they discussing now? Beyond them, the young male driver of the Renault with baby seats got out and strolled over to the skinny trees on the verge. Another piss? Lucky bastard.

'He keeps doing that,' said Charlotte.

'What, having a wee?'

'Yeah, I noticed because I need a wee, but there's nowhere to go. Every time he goes, it makes me want to go even more.'

I hear you, sister.

'He's been about ten times.'

'Ten?' Must be an exaggeration.

'No joke. I've watched him. Must have the world's smallest bladder. He keeps going back and forward. Different tree every time.'

Slack Bladder.

Ten times would be a lot, even for Billy.

'Have you seen any kids in his car?' she asked.

'I assume they're strapped in the car seats. He keeps turning round, must be talking to them, keeping them entertained. I suppose he doesn't want them to get out with this dead body right there. They must be going nuts, though, the car keeps rocking.'

Billy had noticed the rocking too.

She decided to take a look for herself. But first, she took Charlotte back down the carriageway to Mandeep, who sweet-talked his controller into contacting the babysitter. Billy decided to let the stolen vehicle go for the moment; it seemed to be part of some complicated domestic, and she couldn't see how it connected to the dead driver. She would keep an eye on Charlotte though.

She left her with Mandeep and headed in the direction of traffic to the rocking Renault. It had given her an idea.

Years ago, one of her young constables went undercover, doing covert work. She popped back into the nick for a cuppa one day, and Billy recalled her saying that she'd been told by another female officer to always carry nappies in her bag in case she needed a wee on a stakeout.

Maybe Billy could find a nappy?

Car seats meant young children, and young children meant nappies.

She couldn't work out if a nappy would be better or worse than going on the grass verge.

Was she going mad? Had excess liquid flooded her brain? No, she could literally wee into a nappy, that's what they were for. Maybe she'd get two nappies, just in case.

Billy passed Charlotte's stolen SUV and the mother and daughter's minivan – the girl, Hyacinth, was sucking her thumb again – and the rest of the grid until she reached the Renault with its child seats. Slack Bladder was back behind his wheel. Billy glanced inside. No kids. Just two empty child seats.

That meant no nappies.

But also – who had he been talking to all this time?

Chapter Eighteen

6.47 p.m.

The family vehicle driven by Slack Bladder was the inverse of Charlotte's SUV. The Renault smelt like an old biscuit tin.

Billy leant on the open window of the passenger side.

Another tang explained why the driver had been coming and going; maybe he didn't have a weak bladder – he'd been smoking cannabis. Not inside the vehicle, or the smell would have been stronger, but a sweetness was coming off his clothes.

Billy glanced into the back. Two child seats. Both empty.

'You're a young dad,' she said. 'What's your name?'

'Charlie.'

He leant hard up against his door, awkwardly angling himself away from Billy as though she might be contagious. Or about to jump him.

'Do you think that vehicle could be faulty?' Charlie said.

'What?'

'The dead bloke. His sedan. Carbon monoxide could have backed up into the car from the engine and killed him. Through the air conditioning or something. I can smell fumes.'

'You can smell weed.'

'I thought the fumes might explain what's going on. There is no killer. He got gassed, like in an oven.'

Don't give up the day job, Sherlock.

'Do these car seats belong to your kids?' she asked.

'Nephews,' he said.

'This isn't your car?'

'My sister's. I'm on my way to pick them up, we're going to a festival.'

'Does she know you drive stoned?'

'I'm not.'

'I can smell it on you.'

'Prove it.'

'I can't, but I can have you taken down the station and they'll prove it.'

Charlie made a great show of looking up and down the carriage-way. 'Who will take me to the station, then?' he asked, innocently.

'This place will be crawling with coppers soon.'

'Ain't nobody coming any time soon. And you can breathalyse me. I've taken nothing.'

'I'll be contacting your sister, make sure she knows.'

'You do your civic duty, love.' He shuffled his feet in the footwell, regarding them intently.

Charlie was as scruffy as his vehicle. The yeasty smell might be coming from his hair, which was slightly matted at the roots. He wore a cheap hoodie that had gone shiny at the elbows like an old man's suit. His hands on the wheel were tanned and bony, but his nails were long and clean.

'So what do you do for a living?' she asked.

'Brickie,' he said.

'Not with those hands, you're not.'

'What?' He turned his chin in irritation.

'Your hands,' Billy explained, 'are not the hands of a manual labourer.'

'What do you want?'

'I want to know why you keep going into the trees over there. Why you keep making this car rock about all over the place.' She stopped. Charlotte had said he kept turning round towards the back. What was in the back?

'I was looking up and down the carriageway,' he said. 'Curious, wasn't I? What with that dead bloke and all right behind me. I keep turning round to see what's going on. But I don't like to get in your way so I stay in the car. So he's dead, then, this bloke?'

He'd started talking at a rate of knots as soon as Billy looked at the back seat.

Charlie boy.

What are you trying to distract me from?

'I'm going to take a look inside the car, okay?' she said.

'What for?'

She opened his back door. He hadn't got the nous to refuse her permission to search his car. Even if he had done, she would have carried on. This was nothing to do with the murder. This was . . . basic. She could smell a petty criminal a mile off, even without the waft of weed. This numpty would be a nice little scalp to prove herself to the audience on the motorway. The unbelievers. Didn't matter if she ruined a petty crime scene, she'd never get him to court for whatever nitwittery was going on in his sister's Renault. Billy just needed to flex her muscles. Show the doubters her authority. She climbed into the back seat, squeezing her bum over one child's chair to sit in the space between.

'What do you think you're doing?' he said.

She tuned out his faux outrage.

What is it you don't want me to see?

Billy felt inside the pockets at the rear of the front seats, shifted onto one buttock to pull open the fold-down armrest but found nothing behind it, then peered into the boot. She reached over and grabbed a duffel bag, but found it contained only clothes and a can of Lynx. Tossed it back again.

Charlie grumbled while scratching both armpits simultan-eously.

She regarded the child seats. They sat there, unnaturally stiff and upright like kids who'd frozen during musical statues.

What are they hiding?

She released the seat belts and eased one chair out of its place.

Which was when Charlie made a run for it.

He was out of the driver's door and off down the carriageway in an instant.

Billy grappled the child seat out of the door ahead of her and fell after it onto the tarmac. She gave chase around the Renault. There was a shout. A few metres up the carriageway, Charlie was wrestling Pat in the narrow space between two cars. Billy caught up and grabbed his arm, twisting it behind his back. When he kept struggling, she gave it a brief painful tug of warning.

'Don't make it worse for yourself.'

'It's my sister's car. The drugs must be hers.'

'I really need a wee,' Billy said. 'So if you want to take the piss, be my guest.'

Billy nodded to Pat and beneath the combined weight of two units, the young guy toppled face first onto the ground. Billy rested her knee on his bent arm, making him cry out.

A pair of handcuffs appeared in her vision.

Billy looked up to see Daisy.

'Use these,' the girl said.

Billy pulled the man to his feet, accepted Pat's help to shove him into the driver's seat of his own Renault, and handcuffed his right hand to his steering wheel. She stepped back, panting hard.

The nurse was rubbing her chest.

'You alright?' Billy asked.

'Kick in the tit,' Pat said.

'I'll add that to his charge sheet,' Billy said.

'It was you!' she protested.

Billy apologised to Pat and then thanked Daisy for her timely assistance. The girl explained breathlessly that she only had the handcuffs because they were a present. Billy said, 'No one's judging', but her mind was already turning it over; why's the girl got handcuffs in her bag?

Charlie aka Slack Bladder

Renault Megane

I was going to tell her what I saw, but fuck her, she hurt my elbow. And I heard what that bloke in the weird van said, the one on the other side of the motorway, she went white as a bag of ket when he said she'd hurt a kiddie.

So she must of done something.

Yeah, she can swivel. I wouldn't do stuff to kids, so where does she get off lecturing? Usual story, one rule for them, another rule for us. I'll take it to the grave, then, what I saw that driver do.

This handcuff hurts, though. Wrist is throbbing. Hurts even more if it twists, better to keep still. Hard to when it's so hot.

What if it is a murder, though, this fella? Murder is a step up from dealing. Maybe I could strike a deal with her? Information for – what do you call it? – immunity, like they give to gangsters and informants. Not that I'd betray one of my own, I'm sound, but that dead bloke is nothing to me. I was just over in the trees, going about my own business, keeping my eyes open like a smart lad, like I do, that's how I see opportunity coming my way. That's how I saw what I saw. Eyes wide open, that's me all over.

That copper did this handcuff too tight. It's shutting off the circulation. My pulse feels massive. Swelling up. Could lose a hand here. And my sister's going to kill me. I can hear her already, *Imagine what Mum and Dad would of said*, so smug that she got their house and I got nothing. Well, hardly anything.

Copper's coming back. Time to see it, say it, sort it.

Chapter Nineteen

6.54 p.m.

Charlie tried to talk the moment she opened his door, but she showed him the hand.

'I am arresting you—'

'I've got something to tell you first,' he yelped.

'Be quiet or I'll have you for resisting arrest.'

'Don't you want to know what I saw? I thought you were investigating? There's a conspiracy. We'll make a deal, you and me, win-win, it'll work out better for us both. I'll tell you everything I saw. And I won't tell anyone what the van driver said about you killing that kiddie—'

Billy held up one palm again. 'Carrots or sticks, Charlie?'

'What?'

'Are you trying to threaten me or negotiate with me? My advice is to pick one and stick to it. A carrot or a stick. Can't do both at the same time.'

In response, he goldfished so she slammed the door in his dozy face.

'What you have to remember' – she raised her voice to be heard through the glass – 'is that I've had a lifetime of pinheads like you. I'm not listening to any more of your guff.'

She marched over to the grass verge where Slack Bladder had been going back and forth. She could just about hear his muted voice inside the car going on about making a deal, quid pro quo, tit for tat. He was the tit, he'd got that much right, but she was buggered if she was going to be the tat. She dug the toe of her shoe into a patch of kicked-up ground behind the scraggly trees. *Nothing more aggravating than a have-a-go criminal, inept little shit, thinks he knows better, full of piss and vinegar but no idea.*

There was something buried in the dirt. She got down and used her fingernails to lift the loose earth and uncover a hoard of plastic baggies.

Charlie had pulled a *Shawshank Redemption* – making regular trips to dispose of the evidence from his car. Must have crapped himself when he realised he was stuck at a crime scene in a car packed full of drugs.

She carried one little bag back to the car and opened the door to wave it in his face.

'No more threats or bribery and definitely no deal. I'm arresting you for possession of Class A drugs.'

The shock of that shut him up for good. While Billy read him his rights, he stared at his hands, one palm massaging the wrist where the handcuff was biting.

'So where were you taking your sister and kids?' she asked. And when he didn't reply, 'A festival, right? So you intended to smuggle the drugs hidden inside the child seats, and then deal. What would have happened if her kids had got hold of these bags? What if they thought they were sherbets and ate the lot? You'd be up for manslaughter.'

Charlie said nothing. He was visibly shaking now. Billy slammed the door on him again. Let him go stale in his own biscuit tin. He had nothing to say that she needed to hear. She labelled the evidence bags even though none of it would be admissible because she'd forced her way into his car. But she'd stopped his daft plan.

She turned around to see eyes on her from across the carriageway. At least she might stand a chance of keeping control of this scene, if they believed she was a force to be reckoned with.

'That was amazing,' said Daisy.

'God, yah, first rate,' agreed Olly.

'Well, I have to thank Pat for detaining him.'

Pat shrugged modestly. 'I was on my way back from having a wee and he ran right into me. I just held on.'

'Good job,' Billy said. 'And where did you manage to have a wee?'

'Over there.' She pointed to the verge, a few metres along. In plain view of the motorway. 'You get used to going in front of people when you're inside. I literally couldn't give a shit who sees me piss.'

Well, Billy did care – and she couldn't lose the authority she'd just regained by squatting in front of everyone – so that was the end of that. She'd just have to hold it. Even if it was starting to hurt.

She turned in time to see Heathcote giving Pat a full top to toe side-eye scan. The nurse spotted it too.

'What you turning your nose up at?' she snapped at him. 'A woman with bodily functions or the possibility of someone getting a second chance?'

'I said nothing,' protested Heathcote.

'Didn't have to. You expect me to get in this car and back up the motorway of regret, do you? Bit late for that, pal. Regret is not going make any difference. But second chances might.'

The landlord raised his palms to the sky.

'Let's not turn on each other,' Billy said. She put one hand on Pat's arm and sent her away towards her Golf. The silver fox retreated of his own accord to the Beamer.

Charlotte turned back to her car too, yellow tin in her hand once again.

In the inside lane, Kerry and her daughter trailed away. Hyacinth held up one of the soothing sweets to show her mum, and Billy heard her ask, 'Do I like these?' Kerry must have decided that she did because the daughter shrugged and popped it into her mouth. No wonder she was so skinny if she had to check everything she ate.

The calm that Billy had envisaged hadn't lasted more than a few minutes. They were all hot and tired and angry, but were they capable of murder?

If so, there hadn't been so much as a crack in the facade. And yet someone had done it.

So what else was this killer capable of?

Daisy Finch was obsessed with every step in the investigation – and she'd been carrying handcuffs. Pat Mackey was an ex-con with a temper. Charlotte McVie was in a stolen car. Nigel Heathcote acted like judge and jury – was he also executioner? Kerry and Hyacinth kept themselves to themselves. The estate agent, Carl Dawes, had a burner phone and the best position for murder. 'Bob' the Van Man was an aggressive piece of work. And even Olly Sims was a bit too helpful . . .

. . . until she'd got close to his camper van. And then Olly turned distinctly unhelpful.

That was a loose thread she could pull.

He was in the camper van now.

Billy walked the short way down the carriageway. Time for a quick chat.

As she approached his side door, Olly started to slide it shut but Billy called out. 'That's a handsome bike.' He was too polite to slam it in her face. She'd worked a case a couple of years back involving stolen mountain bikes. MTBs, as she'd come to know them. She couldn't believe the spec – Kevlar wheels, carbon frames, bespoke livery – and the cost. More than her car. This one looked expensive. 'So you a professional, then? Looks that way. What are you into? Trail? Enduro?'

Olly laughed off the praise. He slid the door back open.

'I wish! No, just a thrill seeker. I'm that bloke the professionals laugh at. All the gear, no idea. But I'm training to be an engineer and I'd love to design bikes, so.' He spun the wheel of the MTB that was held up on its stand. Its spokes blurred.

Bike spokes.

Billy stepped closer. Olly held out a thumb and the spokes drummed against his fingernail, bringing the wheel to a sudden halt. The spokes were black. Metal rods. Thin as a kebab skewer.

Like the metal thing in Dead Driver's neck.

There didn't seem to be any spokes missing from this bike.

'Do you carry out your own repairs?' she asked.

'Ah, yeah, that's all part of the fun.' He laughed at himself. 'Repairs and modifications.'

'What about spokes?'

'Wheel spokes?' His voice was light. 'Sure. That's basic, like, baby bike repairs.'

'Is it hard to remove one?'

'Not really, you just pop off the tyre and release it from inside the wheel rim – why?'

'Are they easy to buy?'

'Sure. Bike shops. Online. I keep spares.'

'You have bike spokes in this van?'

'Yah, I— Hang on.' Olly opened a drawer, lifted out a cardboard container like a very narrow shoebox, and held it out to reveal a lot of black metal skewers. They had bent ends and flat heads exactly like the one in the man's neck. 'You have to make sure the diameter is right, so I carry some for emergencies. Makes you popular with other cyclists.'

'You like to be helpful, don't you?' said Billy.

Olly seemed slightly affronted by this. 'Doesn't everyone?'

She'd met the occasional witness who was a little too helpful. 'I guess I do like to be helpful.' He started putting the box away. 'But I've only lent you a few old ropes and I'd like to get them back when you're done because decent climbing ropes are crazy expensive.'

'Can I borrow one of these spokes?' Billy asked, reaching into her pocket for a fresh nappy bag to cover her hands.

Olly shrugged and reopened the box. 'Sure, but why . . . ?'

'Working on a theory.' She took the black metal rod he held out to her. 'They come in different diameters, you say?'

'These are all the same. To fit my bike. I did have some thinner ones but I gave them away.'

'Gave them to who?'

'A cyclist. In France. She'd crashed on a trail and taken a few out.'

'So not here, then?'

Olly's top lip was curled up now in open confusion, revealing his perfectly trained teeth. 'Look, what's this about? What's with the plastic bags on your hands as though you're handling evidence?'

'The whole carriageway is evidence. You're helping the police with their enquiries and you've been really helpful, thanks.'

He shrugged one shoulder. On the petulant side of polite. He was annoyed but too disciplined to show it. In Billy's experience, people who keep their emotions under control in public are often the ones who go nuclear in private.

Her pace was slower than her pulse as she crossed two lanes and ducked under the cordon to reach the sedan. This could be a breakthrough. She put on fresh nappy bags and opened the back door. Heat swarmed her. She took a fresh breath before leaning inside the stifling cabin where thick air laid itself over her face. The smell of blood and warm urine was concentrated after almost two hours in the heat. The man's neck was paler than before. The thin rod that protruded a centimetre from his nape was visible. Billy held up the curved end of the bike spoke without touching him, comparing it to the curved end of the thing that had killed him.

Same thickness. Same shiny black. Same flat-headed end.

Same.

It was a bike spoke that had killed him.

THE THIRD HOUR

Chapter Twenty

6.59 p.m.

A bike spoke. There were gangs – petty street thugs – who used sharpened bike spokes as cheap, lethal and easily concealed weapons. But this wasn't a case of gang violence. Was it? So how did a bike spoke end up in the back of this driver's neck?

'Detective?' Mandeep and the boy, Jack, approached the cordon. 'I have a message for you. Came through on the radio.'

The taxi driver held out a scrap of paper. A note scribbled on a taxi receipt.

Chad might not be Chad.

'Superintendent Day asked you to call when you can,' said Mandeep.

'What's the time now?' Billy asked rhetorically while checking her own watch.

Mandeep answered anyway: 'Almost seven.'

Two hours she'd been on this motorway, and she was back where she started.

The only hard evidence she *thought* she'd found – the dead man's identity – was possibly fake.

Except . . .

One new thing: who carries fake ID? Kids trying to buy alcohol. Or criminals.

Chad McClusky – *or whatever his name is* – wasn't the former so he must be the latter.

Who are you, Dead Driver, what have you done?

Billy whipped open his passenger door and rummaged in the glove box, ignoring the hot-crotch smell of the interior. Maybe something could shed light on Not-Chad's identity. She found a car rental agreement.

Like her own, it was folded inside a sky-blue paper sleeve. They'd used the same company at the same airport. Maybe they'd even waited in the same line together. His face, his profile, rang no bells.

But then she'd been in a daze at the airport, both from the long flight and from arriving back in England after six months away. Overwhelmed by normality, she found herself hyper-aware of familiar details. She'd noticed things that she'd never previously recognised as *things*; shopfronts, dress codes, signage, the tinny nature of English light, the pall of fast food, the precise amount of personal space that Brits allow one another when they queue as if they've bloody measured it. Accents, scowling, socks. By the time she got to the rental desk, she'd run out of mental bandwidth and failed to notice a slightly handsome American—

Oh, silly me.

If he's not Chad, then he might not be American.

She stood up too quickly, head spinning. Jet lag. Heat. Dehydration. Needed to watch herself. Her vision cleared. She glanced at the Hybrid Welfare Unit on the other side. Her mind had only gone down the CIA route because she thought Dead Driver was American. With that connection gone, the suspicion felt flimsy. Outlandish, even. Any dread had evaporated. The van was just a big, black vehicle with tinted windows.

Now I'm scared of tinted windows.

Ffs, Billy.

It was the same as the menopause-driving-phobia-thing that had side-swiped her in Australia. Where she'd cut across two lanes of traffic – cars swerving, horns blaring – in her panic to get to the hard shoulder and just . . . make it stop.

Hands opening and closing on the wheel like beached fish.

The uncontrollable force of anxiety had taken over her body. Like a She-Hulk.

You won't like me when I'm anxious!

Get a grip, Belinda.

It had started again on the motorway earlier. The same force had started building. Who knows what would have happened if the traffic hadn't stopped when it did. She might have She-Hulked again, and even now be sat further up the hard shoulder, hands grasping at the wheel, crying and wondering how she'd get home without her sister to pick her up this time. Oh, God, the only option would be to call the police. The humiliation of another officer driving her car off the motorway – some smug baby PC who thought her knees and self-confidence would last forever.

Billy pressed her fingertips into her left boob, feeling the scuttle of her heart.

And now this stupid black van with tinted windows.

How could she go on working if she couldn't tell the difference between the gut-feeling of intuition that had always served her well as a copper, and this new clanging bell of anxiety that was irrational and misleading? How could she return to policing if she couldn't judge what was real and what was inside her head?

Maybe she should speak to Dom on the taxi radio . . . Hear a friendly voice.

But if she was going to call in, it would be better if she had something new to tell him. Otherwise he'd just think she was losing it. Again.

So, Dead Driver. *Who is he?*

She ripped open the envelope of car rental papers. They were signed Chad McClusky and the given address was Howle Green.

She glanced over at Heathcote, the landlord from Howle Green. He was arranged like some kind of Rodin sculpture, leaning against his car with chin propped on clenched fist. *The Thinker*. What was the silver fox thinking so hard about?

Oh, silly me again.

The Howle Green connection might be fake too. Perhaps Not-Chad had given a fake address to go with a fake ID. It would explain why Heathcote hadn't heard of the Airbnb near his pub.

Christ, I really am starting all over again.

Two hours and I know nothing.

What *did* she know about Dead Driver? Not-Chad had arrived on a flight from Washington on false papers. The Washington connection was a real lead. If she could confirm that, it would be a start . . .

What was it Pat said earlier? When the authorities in India blocked the local network, the international phone lines still worked. Okay, it was worth a try. Billy dialled the international operator and it connected, rang. When the operator answered, she asked to be connected to Dulles International Airport in Washington.

An unfamiliar dial tone rang five times before it was answered. Billy asked to speak to security.

'What's it regarding, ma'am?' The woman sounded officious even from thousands of miles away.

Billy gave her credentials and explained that there was an ongoing terrorist incident in England, which meant they needed urgent information about a passenger who left the airport this morning.

'There are official channels for enquiries, ma'am. It's irregular for first responders to call on the public line.'

'This is an irregular situation,' Billy said. She'd hoped the operator would hot-potato the call to a security officer, and she could appeal to them law-enforcer to law-enforcer. 'Local domestic phone lines in the UK are down right now so I'm unable to go through the regular channels. We're in the middle of an incident involving mass fatalities.' Billy could hear the clack of false fingernails on a keyboard; a second later the woman cussed softly under her breath. Billy tried to drive home the advantage: 'It's imperative that we identify a suspect who flew into our city from Washington right before the attack.'

'You still have to go through regular channels, ma'am. I'm sorry I can't be more help—'

'Please don't hang up! Okay, I understand that you have your protocols, so could you please ask your head of security to contact British law enforcement via the official channels? I can give you a name. My senior officer has security clearance.'

A micro-sigh. 'What's the name and number?'

'Superintendent Dominic Day.' Billy gave the rest of the details. The operator hung up after a warning that she was making no promises.

Billy felt something inside her shift. In the great Tetris game of an investigation, one awkwardly shaped piece was now lined up and inching closer. She just hoped it would drop into place.

A yelp on the other side made her shoulders snap up. A mother raising her voice to a child – *I don't know how much longer!* – and the kid ugly-crying in response.

Nerves were fraying. How many acts of violence had Billy responded to over the years that were caused by frayed nerves? Countless. Frayed nerves, she could understand, inexplicable acts borne of high emotion. But this killing felt different. The bike spoke suggested this was hatred; it was pure, premeditated, and pitiless.

She had to work faster.

The murder weapon. That was real progress. It meant her instincts could still be trusted.

Sharpening a spoke into a lethal spike . . . What did that say about the killer?

They'd had to conceal the weapon.

They weren't afraid of getting their hands bloody.

So they must be angry.

They'd had to get the job done fast – *jab, jab, jab*.

And they could still be here in plain sight. Literally any of these people could have a weapon up their sleeve.

Chapter Twenty-One

7.05 p.m.

Billy became aware of a woman waving a white cloth over the central reservation. She was either trying to surrender or she wanted Billy's attention.

On her hip, the toddler who'd been ugly-crying only minutes ago was listless. The child had a stripy Breton-style top and sweat-damp blond curls around pink ears. A boy, Billy noted as she crossed the outside lane to reach the woman, who almost certainly had a very middle-class name.

'This is my son.' The mother made it sound as though this was Billy's fault. 'This is Otto.'

'Hello, Otto,' said Billy.

'Otter,' said the woman. 'His name is Otter.'

Yup.

The rabbit hole loomed again. This was one of her bitter little bugbears – people who are so blithe about parenthood that they treat their children like fashion accessories.

Focus, Belinda.

She took the boy's pudgy damp hand and jiggled it. 'Sorry about that, Otter.'

'Well, I need to see about getting us off this motorway, because this little boy has had enough. It'll be getting dark soon, and I simply don't think it's safe. There's talk of killers and car bombs and now drug addicts. It's out of control, frankly. I'm concerned about Otter's physical and mental well-being.'

Ooh, mental health.

Playing the trump card in the first hand.

Otter grabbed the neckline of his mother's top to swing himself around to face Billy, treating her to a flash of Mum's deep and sweaty cleavage. The woman snatched the fabric back up.

'I do understand your concerns,' said Billy, 'but the safest option is to wait inside your car. You can lock the doors if you prefer. And if Otter is struggling with the heat—'

'Otter is fine in warm weather. He's a very frequent traveller. That is not the problem here.'

Billy tried again: 'If you need more water—'

'I always travel with drinks and snacks. What we need is to get off this motorway.'

'That's impossible until the traffic moves. The hard shoulder must stay clear for emergency vehicles.'

'What emergency vehicles? We've been abandoned. It's an outrage, frankly. There are children here.'

'There's nowhere to go even if you drive away. The connecting roads are also gridlocked. We are going to have to stay calm and wait it out. If you need any help—'

'I don't need help, I need assistance. There's a difference.'

'Is there?' said Pat from behind Billy. 'Why don't you explain that difference?'

'Well,' said the mother. Otter tugged her top down again and she slapped his hand. '*Help* implies I'm in distress when, in fact, I'm simply asking someone to do their job properly. Which, in this case, means offering *assistance* to those most in need as a first recourse.' She flicked up her hip and Otter suffered minor whiplash as he settled more securely in place. 'I think we can all agree that Otter deserves to be kept safe.'

'Which she's already told you to do by sitting in your car and locking the doors,' said Pat.

'And who are you?' asked the mother.

'My name's Pat. I just got out of prison after twelve years, so I can handle myself. If you like, I can come over and sit in your car, keep you safe? Happy to offer my full assistance.'

'I'll be filing a police complaint against you,' said the mother with a finger-jab at Billy. She kept up the complaints even as she walked away with Otter staring over her shoulder. 'You have endangered an innocent child with your attitude. We should be evacuated . . .'

'I could administer an enema,' muttered Pat.

'In fairness, she's not wrong,' said Billy. 'It's going to be hard to maintain order when it gets dark.'

'So you'd better crack on, hadn't you?' said Pat.

Billy refocused her mind on the murder weapon.

The bike spoke that rather pointed at Olly as the main suspect.

But this killer had put the death of Dead Driver above all else – even the instinct for self-preservation – because, realistically, he couldn't get away with it. Could he? A murder in broad daylight in a public place. This killer's thread of humanity, maybe even reality, was as thin as a bike spoke. Not unlike the terrorists in the city, prepared to give up their lives for a cause. Fanatical. Was she dealing with a killer or a zealot?

Did Olly seem like either of those types?

No.

But he had lured her away from the crime scene. Olly was the one who told the drivers near the abandoned car that she was a copper. That meant she had to go up there and deal with the 'car bomb'; a lengthy distraction, as it turned out. Did he do that on purpose to stall for time? If so, he was manipulative, opportunistic, able to think on his feet.

Don't underestimate an affable posh boy who needed a haircut. Had British politics taught her nothing?

'Knock knock,' she said as she reached the side door of his camper van. Daisy sat in the back with her arms wrapped around her bent-up knees. 'Can I have a word, please? Alone?'

The girl unfolded herself and left them to it.

'Do you know the name Chad McClusky?' Billy asked.

'No,' said Olly.

'Take your time.'

'It's a distinctive name, I don't know it.'

'Do you recognise the deceased man? Ever seen him before?'

'God, no, why?'

'Would you object to me searching your van?'

'Don't you need a warrant?'

'Not if you give me permission.'

'My mother's a lawyer. A criminal barrister, actually—'

Oh joy.

'I could phone her and find out if you've got a leg to stand on, but I strongly suspect you are, in the current circumstances, utterly legless.'

'Feel free to call your mum once the phone lines are reconnected. I'm not infringing anybody's rights. I'm asking nicely. But as you will have noticed, I'm in the middle of a murder investigation and the clock is ticking.'

'Not being funny, but I would have thought the car bomb is the priority?'

'There isn't a car bomb. Not here. Those explosions were just a warm-up for the bigger incident in the city centre. So my priority is this dead driver.'

'What does that have to do with my camper van? I've tried to be helpful. I've been your errand boy, lent you my gear, which I hope to get back, by the way. And all you've done is victimise me.' He prattled on some more about his mother, the criminal barrister.

Billy took a deep breath and released it in a slow stream through her nose. As his energy raced upwards, she had to lower hers to provide a counterbalance. Sure enough, after a few more moments of bluster, he ran out of puff.

'Alright, fine!' Olly leapt out of the van, shoulder-barging Billy as he passed. He spun round on the tarmac to face her. 'Everyone thinks you're not a real copper, you know that? We don't have to let you go through our stuff. But go ahead, turn the van upside down.' He scratched frantically at the top of his head.

Billy's sister had had a tic like that when she was young. She'd make herself bleed at exam time. Once upon a time, Mel had been the highly strung one, Billy had been rock steady. But then Mel hadn't had thirty years of people giving her asshattitude.

'It's not like the movies,' said Billy. 'I don't need to turn your van upside down.' She checked that her makeshift gloves were in place. 'I'll go carefully.'

Olly said nothing but scrunched his lips up under his nose like a little boy. Billy wondered what on earth reason he could possibly have to kill the man in the black sedan but climbed into the confined space of his van anyway.

She worked quickly, taking photos as she searched. She squeezed around the bike to check a cabinet. What was she looking for? Actual sharpened bike spokes ... Or anything that connected him to the dead driver, whatever his real name might be.

Handcuffs.

In the top left drawer of the cabinet, Billy spotted a pair of handcuffs. She took a picture before holding them up to show Olly.

'Why have you got these?' she asked.

His eye roll spread to his shoulders.

'Just a joke, someone gave them to me.'

Feasible, but still. Handcuffs.

Daisy had had a set too that were 'a present', now these ones were 'a joke'. Is this a Gen Z thing? Maybe they had to be Tinder-ready at all times? She decided to confiscate the cuffs. If nothing else, she wanted to take them out of circulation while these people were all trapped together on a motorway. The key was there too so Billy dropped both parts into a poo bag and tied it up.

The other drawers contained only tools and accessories that looked to her untrained eye like the real paraphernalia of an adrenaline junkie; carabiners, WD-40, tiny screwdrivers. Maps. Gas for a camping stove. A camping stove. Packets of soup, mugs, pans. Those implements that are half spoon, half fork, whatever you call them.

No sharpened bike spokes sitting there like a smoking gun. Billy sat back on her heels.

Sporks, that's what you call them.

How could Olly have murdered Dead Driver anyway? He would have had to run across all three lanes of the carriageway – outside lane to inside lane and back again – without being seen.

Unless ...

Was there anywhere for a second person to hide?

Billy lifted the cushions on the benches and the lids of the wooden seats beneath, but the storage boxes were filled with gear, not a hidden murderer.

The cardboard box of normal bike spokes was on the side. Billy opened it and ran a gloved fingertip through the rods. None was sharpened.

'Ah, you might want this?' Olly held an envelope stuffed with receipts. He rifled through them and held one out. 'Receipt for twenty bike spokes. Bought them last week. They're all there, bar the one I gave you before. Just in case you thought one might be missing.'

'Why do you think I might be looking for a bike spoke in particular?'

'Because you're staring at them? Like, really intently? And taking photos. And you took one off me before. And you're investigating a murder. And you clearly think I'm an entitled posh idiot, but I'm also the son of a criminal barrister, so I'm not a completely ignorant entitled posh idiot.'

Which was exactly what Billy had been thinking. She counted the spokes in the box and there were, indeed, nineteen plus the one she'd taken before that was still in her car. It didn't prove anything, but she made him drop his receipt into a poo bag. She inspected it through the thin plastic, mostly for show because it was in French, which she did not speak. Olly almost certainly did speak French because he'd had that kind of schooling, and she tried not to be bugged by a gnat of inferiority because she needed to focus now.

'Is it a long drive to Corsica?' She pointed at the address on the receipt.

'Couple of days down to Marseilles, overnight ferry. It's not so bad.' Olly patted the camper van like a horse.

'Alright,' Billy grunted as she moved around the mountain bike to the exit. God knows why people liked camper vans. More work than rest. She put one hand on the driver's seat to support her weight as she prepared to step down onto the road. Caught sight of a mobile phone in a holder on the dashboard. Olly must have been using it for the GPS. She leant between the seats and tapped the screen; it went straight to Google Maps. Billy used two fingers to pinch it out, to see the destination on the screen.

She zoomed in on the pin. *Wheelers Supplies.* Fair enough. He was on a detour to get bike parts, just as he'd said. She nodded at him as she got out.

'Am I free to go?' He smiled with one side of his mouth.

Billy matched his smile, glancing around the carriageway to check that everyone was where they should be. Daisy stood close enough to listen in. She pulled herself to attention and started asking about the handling of evidence. Chain of custody, what an officer decides to confiscate and how it should be labelled and how she would ensure it wasn't contaminated or tampered—

'Today is far from textbook,' Billy cut in.

Daisy nodded, colour high on her cheeks as though she found the whole thing thrilling. What did she say she was studying? Criminology? Politics? Something like that. Maybe Billy would check that her studies were the only cause of this close interest. The girl had thrown on a ragged straw hat. It had a seahorse on the front and the word *Corse*. Even Billy knew that meant Corsica, mainly because she'd just read it on Olly's receipt.

The pair had said they met here on the motorway. And maybe Olly had lent her the hat. But there was some connection between Daisy and Olly that scratched the back of Billy's nose like a sneeze that wouldn't come.

Oliver aka Byron

VW Camper van T2 Split Screen

My hands are rock steady. She's just a sergeant, not senior enough to be a bother. She knows nothing. I'm fine. Run out of disinfectant wipes, though, I'll have to unpack the whole van and put it through the dishwasher when I get to Mum's, must be marks on everything as though one of those UV lights is showing up blood or semen or whatever, germs and bacteria, and the last thing the sergeant touched was a dead body so there's that as well, even though she was wearing bags over her hands it feels like her sweaty fingerprints are on everything including my face and lips.

There might be disinfectant wipes in the drawer?

No.

I slam the drawer with my thumb in it. Sit a while on the passenger seat to nurse the pain. It throbs like a stunned bird against my chest. Humble little sparrow.

Daisy is standing beside her hot-pink Micra. Very flirtatious colour. The blokes from the other side of the carriageway were sniffing around but they lost interest when she got officious on them. They thought she was a fun girl, in that shimmery dress, turns out she's serious. My straw hat looks good on her. But she'll have to keep it now. Dead skin. She's unbelievably beautiful, the sun ignites her long hair. I'll wait five minutes and go over.

Need to clean the van, first. Thumb seems fine.

Maybe under the seat?

Yes, spare packet of wipes. Result. Start with the bike spokes which the sergeant had in her filthy hands, gloves or no gloves. It'll take ages to clean this lot, more than five minutes, but I'll check on Daisy as soon as it's done.

Shame about the handcuffs, should have stood my ground, not let the woman dominate me, I need to stand up for myself, that's what Mum says, so exasperated with the flakiness, and it's true, I am bloody useless at anything that requires a bit of application. Like Mother, Daisy is going places. Criminology and Politics. Serious stuff. Where do I get off thinking I could be with her, entitled little shit that I am, it's the privilege talking, none of it earned. The trapped tears will only give me a headache so I let them fall into a disinfectant wipe. Then I'm done.

I nudge my screen to check WhatsApp. A message from my mother sent before the signal went down requesting my ETA for supper. She'll be worried. She'll see the two ticks, know that I've read it but won't know why I can't reply. I type a row of van emojis and a weeping face. *Traffic jam.* Then I delete it because it's infantile. She says emojis are for the inarticulate. The phones are down anyway. Let her wait.

Chapter Twenty-Two

7.10 p.m.

Billy caught her own reflection in the window of Daisy's pink Micra as she passed. Even faded and translucent, she looked exhausted. She shifted her focus past her own face to glance inside the car. It looked like someone had tipped a suitcase onto the back seat. How could the young woman's mind be so sharp and her car so messy?

On top of the mound of clothing lay a T-shirt with a picture of a mountain bike. And a date – the previous Friday – and the word Corse again. The shirt came from a competition or a race only last week. Why were they saying they'd only just met? This doubt had been niggling at her for too long.

She turned to face Daisy.

'Tell me about Corsica,' Billy said. 'You were there last week.'

The girl hesitated, adjusting her straw hat. Olly emerged from the camper van holding a packet of wet wipes and sauntered to her side.

'Yah, crazy coincidence about Corsica.' He nudged Daisy to agree. The girl grabbed her biceps as though he'd given her a dead arm.

'I don't know what's going on with you two, but I don't like lies or omissions,' said Billy. Maybe it was jet lag, or needing a wee, or the horror of seeing herself in a reflective surface, but she felt too disjointed to tie up her words in pretty bows.

There was a significant look between the two youngsters, then Daisy opened her mouth to speak—

As Billy's phone started ringing.

It had been such a long time since anyone's phone rang that it silenced them all.

Number withheld. It must be the airport in Washington, an international call working despite the local network being blocked.

Billy walked away to the hard shoulder before answering. She gave her full name and title.

'Ma'am, my name is John Vega. Airport security. I understand you have a situation?'

'Thank you for calling me back.' She glanced at the sky to offer thanks to the operator who took the initiative to get a message through to this guy.

'We've got clearance to speak, but I must say this is pretty irregular.'

'I agree, sir. I'm having a very irregular day. I am standing on a gridlocked motorway with a man who has been murdered inside his vehicle. And I need to find out who killed him because that person is still at large.'

'How do you know the perp is on the scene?'

Billy carefully explained the layout of the motorway; the high barrier and blocked tunnel that trapped them here, the fact that no one could leave in a vehicle or even on foot. Not without being seen.

'And literally the only thing I know for sure,' she concluded, 'is that this man boarded a flight from your airport this morning to come here. He's now dead. And there is a major terrorist incident in the city only a few miles away.'

'I'm watching it on my screen, ma'am.'

'And I don't know if these events fit together in some way. Maybe they don't. But if there is anything you can tell me about this man—'

'I'll do what I can. Sergeant Kidd, is it?'

'Belinda Kidd, yes. But everyone calls me Billy.'

'Of course they do. Call me John. Do you have a name for the passenger? Flight number?'

Billy told him the name and that it was probably fake.

'And a description?' Vega asked. 'So I can track him down on the security cameras?'

Billy regarded Dead Driver through the windshield. She hadn't looked at him this closely. Now, she opened the door and peered at him. 'So he's got dark hair that's short at the back but quite long on top.'

'Can I assume we're talking about a Caucasian male, as you didn't stipulate otherwise?'

Billy winced. 'Yes, sorry. He's a white male, I'd estimate around the age of thirty. Handsome and well turned out.'

'Sounds like a catch. His clothing?'

'He's wearing dark blue trousers that aren't jeans, but aren't suit trousers either.' She rubbed the fabric between finger and thumb, feeling its density even through the plastic bags covering her hands. 'They're like jogging bottoms.'

'Sweatpants?'

'That kind of fabric but smart. They look new or they've been dry cleaned. They have creases across the knees from the hanger.' Billy looked down at her own rumpled jeans and was suddenly aware of a sharp tinge of her own odour. 'His clothes look pristine, not like they would off a flight.'

'Maybe he changed? Sounds like he could have purchased new clothes?'

'Maybe.' Billy asked, 'I guess a lot of people freshen up after a flight?'

'We have showers you can rent by the hour. I wouldn't say that's unusual.'

'So let's assume his clothing might have changed. You're looking for a white male, dark brown hair that's cut short at the back and sides, but floppy on top.'

'Floppy?'

'Think Hugh Grant in *Notting Hill*. Less floppy than Hugh Grant, obviously.'

'Got it. Height?'

'He's sitting in his vehicle, so it's hard to be sure, but he looks to be average height.' She peered down into the footwell to gauge the length of his legs. 'He's got distinctive shoes. They're trainers.'

'Sneakers?'

'Yeah, but high up the ankle. Leather. Nike brand. They look new, clean as a whistle, but kind of old-fashioned.'

'Do they have the jumping basketball guy on the tongue?' Vega had a slight smile in his voice.

'Yep, and a kind of crest, like . . . I think it's the propeller of a plane?'

A pause while Vega thought about it. 'I don't think it's a propeller. It's a basketball.'

Billy peered closer. 'So it is.'

'Vintage Air Jordans. Guy's got class. Okay, I'm going to take a look on the security cameras.'

'It would be good to know if he was travelling alone. How he got to the airport – the numberplate of a cabbie who might be able to give us a pick-up address, or a drop off by a wife whose plate we can run. Any clues to his real identity. Anything suspicious, obviously. And' – Billy sighed at Not-Chad's peaceful face – 'anything that might tell us if he's British or American and why he got himself murdered on an English motorway.'

'Can I ask how he was killed?' Vega said.

Billy glanced over her shoulder to check that no one was listening in. 'He's got a sharpened bicycle wheel spoke in the back of his neck.'

'Oh.'

'Yeah.'

'Brutal.'

'It was. Premeditated. He landed in the UK and was dead an hour later. It was an assassination.'

A beat passed while this landed. It was the first time that had occurred to Billy.

Assassination. Her eyes slid over the rail to the CIA van. She'd discounted it, but maybe too soon.

'So I think he must be a US national,' said Vega pensively.

'How so?'

'I can't find him registered on the FRT.'

'FRT?'

'Facial recognition technology. Okay, let me tell you how it works. When you come through the airport, there are various checkpoints. In the bad old days, you had to hand over paperwork at each stage; your airline tickets, your passport, your boarding pass. But now we've boosted the number of touchless terminals, which means we use facial scans in place of physical documents. Are you with me?'

'I am.'

'This makes the airport smoother and safer, but – there's always a catch – there's a significant portion of the population who object to the gathering of biometric data. Even if it's only to keep them safe, right?'

'Right.'

'So there are ways to legally opt out of the facial scans while passing through the various airport controls. However . . .' Vega liked dramatic pauses.

'Yes?'

'That's only for US citizens. Foreign nationals cannot opt out. They must use the FRT. So if we don't have him registered on the FRT then he must have opted out so he must be American. Do you follow?'

'No FRT, no foreigner.'

'Right. All I can say without further investigation is that he travelled on a US passport today otherwise I'd be looking at his face right now.'

'Right-o.' Billy fell silent.

'You were hoping for more?' said Vega.

Billy's body sucked in a gulp of air. She gathered herself and asked to take his direct line. She carefully typed it into her contacts as she strolled back to her own vehicle. She must have sighed because he asked how she was doing. Pretty stressful situation she was in right now . . .

'I was on my way home from a holiday,' she said. 'About to retire.'

'Ain't that always the way?' Vega gave a puff of laughter, which felt strangely intimate despite the distance. 'But retire? I don't think so, Sergeant Billy the Kidd. Sounds like you got some fight left in you yet.'

Chapter Twenty-Three

7.18 p.m.

Right. Another Tetris block lined up. She'd have to wait for John Vega to drop into place. Now it was time to flip Daisy and Olly around and see where they fit into the picture. The pair were standing by his camper van. She finished her bottle of water. Watching her drink made Billy feel parched; how was it possible to be simultaneously thirsty and needing a wee? And how come the girl didn't need a wee? Skinny little thing like her, where did she put 500ml of liquid?

'You don't seem old enough.' Pat appeared by Billy's side, holding a tiny electric fan close to her mottled neck. 'For retirement, I mean.'

Woman popped up like a genie.

'I'm not really. I just turned fifty-one. But I've been in the force since I was nineteen, so I'm eligible to retire now if I want.'

Pat gave a downward smile of approval.

'But I've been stalling.' Billy stopped. Pat was like human alcohol, so easily did she get people talking.

'I don't think you should retire,' said Daisy from a car-length further along the carriageway.

They've all got a bloody opinion.

Or the girl is trying to butter me up.

'The way you saw through that drongo—' said Daisy.

'What the hell is a drongo?' Olly burst out laughing. 'I have literally never heard you use that word before.'

Interesting.

Daisy flapped her hands. 'Australian soaps! My guilty pleasure.'

Before Billy could speak, Pat pounced. 'I thought you two had only just met?' The nurse was displeased, hands on hips.

'They haven't,' said Billy. And then to the young couple, 'Quiet word. On the hard shoulder, now, please.'

Olly and Daisy did a walk of shame to a patch of sad grass beyond Dead Driver.

'Do we have to stand here?' asked Daisy. 'Close to him? It's distracting.'

Billy directed them to the space between the black sedan and the Range Rover containing the estate agent, Carl, with his drawn-on-with-a-Sharpie beard.

'Here,' said Billy, turning her back to him. The young pair dutifully stopped.

'We already know each other,' said Daisy suddenly.

'Daisy!' protested Olly.

'I love you, but I'm not going to prison for you!'

'No one is going to prison.' Billy shook her head. 'Just tell me what's going on. I really don't have the bandwidth . . .'

'You'll have to say it,' said Daisy to Olly. 'I didn't want to lie in the first place. Now it's just worse. This could ruin my career!'

'I can't say it, it's embarrassing.'

'Well, the whole thing was your idea. So you—'

'Kids,' said Billy, holding up one hand.

They both shut up.

'When I say that I have seen and heard everything imaginable, I mean literally every embarrassing thing. I have seen a woman with her head wedged inside an Aga oven. I have seen a man who choked to death on an inflatable doll. I have seen another man who handcuffed himself to a horse.' Billy stopped. 'Oh, I see. The handcuffs. You had one set each.' Daisy's inappropriate dress. Olly's puppy energy. She'd thought he was just posh and smug, but now she saw why he was so pleased with himself . . .

Two sets of handcuffs. One each.

This was some kind of sex game.

'Tell me,' said Billy.

'It's not like we're into dogging,' said Olly in a tone of voice that his mother might use if she was accused of not scanning an item at the self-service checkout in Waitrose.

'You're too young and attractive for dogging. It'll be years until you're that desperate. So, what are you doing?'

Daisy gave a great huff of acquiescence that swung her pretty hair. 'We just pretend we don't know each other and meet in public and, you know, act like we're strangers.'

Ah, bless.

'We use fake names and—' said Olly.

'Too much information,' hissed Daisy.

'So you two young things spend your days driving up and down motorways, dressed up to the nines, waiting to bump into each other – and then. Where? At a motorway service station?'

'Yah.' Olly's eyes narrowed as though he feared saying the wrong thing and getting himself into trouble. 'Or a lay-by or a country lane. Somewhere we can't be seen.'

'And then—? Oh, in the camper van, obviously.'

Daisy shrugged.

'So long as you're not in public,' said Billy.

'Oh, no.' Daisy shook her abundant hair emphatically.

'Oh, God, no,' Olly furiously agreed.

'So why the big play act?' Billy asked. 'Pretending you don't know each other.'

Daisy rolled her big blue eyes. 'Because when the traffic first stopped, he sent me this.' She woke up her phone and held up a text message for Billy to see. It comprised three emojis; a van, a mustard-coloured woman disguised by a hat and dark glasses, and a rocket. Daisy lowered the phone. 'So we were going to play—'

'Right here in the traffic jam?' clarified Billy.

'It seemed like we'd be stuck for a while,' said Olly.

'And so we started pretending we'd only just met. Like we do. But then the bloke turned up dead.'

'Passion killer,' said Billy.

'It seemed inappropriate to continue with the Game,' agreed Daisy. 'And we thought it would look suspicious if we changed our story.'

'And you seem to have it in for me, quite honestly, even though I've been nothing but helpful,' said Olly. 'So I didn't want to give you more ammunition.'

'And Corsica?' asked Billy.

'We spent half-term there. Just got back.'

'This is my mum's car,' said Daisy, indicating the pink Micra. 'I borrowed it for the day.'

'To go not-dogging?'

'Her mum doesn't have to know, does she?' said Olly. 'Only she reupholstered the seats in my camper van. It seems a bit disrespectful to her cushions . . .'

A shout from behind saved Billy. It was Mandeep, from the taxi, calling down the carriageway from a respectful distance.

'Your boss on the radio,' he shouted. 'The detective.'

Not a detective. Not my bloody boss.

But she raised a hand to signal she was coming.

'If that van starts rocking' – Billy warned Daisy and Olly – 'the police will come a-knocking.'

Daisy aka Mermaid

Nissan Micra

His eyes stay on me all the way to my car. Feels like bird mess oozing down my back. I'll suffer for taking this moment to myself but I have to walk away otherwise I might say something I regret.

The Micra is a sauna. The dark little cave of the footwell looks appealing, I could curl up down there on a pile of shoes like a pet dog. Instead, I drag a stray top from the passenger seat to cover my bare legs. My trackie bottoms are in the mess in the back where he chucked all my stuff out of his van but if I turn around to look for them he'll be watching and if I make eye contact he'll take it as an invitation and if I avoid his eye he'll sulk in the van and I'll have to deal with tears. Tears and talking. It's endless. The only thing that could make him feel better is if I shrunk down to the size of a Funko pop and he stuck me on his dashboard so my head could nod at him in constant agreement.

I only play the Game, he says, *because you're so perfect that everyone wants you and there's a lot of temptation unless I keep you happy.* And I tell him I am happy. He's handsome and sexy and generous and everyone says we're perfect, even our parents say it, but they don't live with the constant drama. The emojis. I have to reply as soon as they arrive or else he checks the ticks and asks why I'm ignoring him or why I'm not on my phone and did I not check my phone because I was too busy and what was I too busy with? Was it someone else, is there someone else? And even if I say I was only chatting to Jess or Aisha he'll say, *But they're single and go out on the pull every night* – which they do, to be fair – and he'll say *Maybe you'd prefer to be out with them, on the pull*, which comes back round to the Game and how he has to do it to keep me faithful. Then he cries. And I apologise. Don't know what I'm apologising for half the time. Even now the phantom guilt is gnawing my gut. Something nasty inside me that must be to blame. It's the crying that does it. I've literally never seen a man cry IRL. Not my dad or my brothers. Except over football. So I must've broken him. Must've done something to make him feel bad, bad enough to cry, and I feel guilty even though I didn't check my WhatsApp because I was at the gym and I haven't spoken to Jess and Aisha in ages because they've given up on me, and who can blame them? I act like I'm married to him or something, worse than married, because Mum's married and she can see her mates without Dad crying.

It's almost like—

Being imprisoned.

Handcuffed.

But then again, Jess dated that bloke who hit her with a spatula.

What if I end up with someone like that?

That's worse than crying, isn't it?

So I'll get my handcuffs back off Sergeant Kidd, give them to him as a present, that'll keep him happy. And if I twist my arm right back in its socket and feel around on the back seat, I might feel the laces from the waistband of my trackies. Even if they're too hot, I can pull them on and anything's better than this fuck-me dress, especially now everyone out there knows about the Game—

The phone pings. But the phones are cut off, how is he—

Pings again.

He's AirDropping me.

Ping.

It's like being strangled without hands.

Ping. Ping. Ping. And I'm gagging on tears that I can't let out, that would really upset him, he can't cope with me being down. I put my bare foot on the accelerator and pump it against the floor. Pump pump pump. *Ping ping ping.* Pump pump pump. And then I keep going until I'm jerking about like a crash test dummy. Back of my skull hitting the headrest, hair crazy all over my face, arms having a fit. The mentalness of it is exhilarating. In control and out of control all at the same time. Pump pump pump. The relief.

So then I stop.

The phone has stopped pinging. Probably, he can see. He knows.

It's tempting to floor the accelerator for real and drive and drive and keep on driving until the road or the petrol runs out, whatever comes first, and just stay there. I know I'll never find anyone who loves me as much as he does. But there are moments when I wonder if it might be better if he loved me a little bit less.

I shake my hair off my face. Peel away strands that have stuck. Use the fingertips of both hands to press the skin under the eyes, make sure they're not puffy. Have a good sniff. Then I get out of the car and go to reassure him.

Chapter Twenty-Four

7.22 p.m.

As Billy walked down the carriageway towards Mandeep's taxi, she dialled the number in Washington, praying that the international lines still worked even if nothing else did. But her relief as it connected faded as the phone rang about ten times. Finally, it was answered by a woman who simply said, 'What?'

Billy asked for John Vega.

Hand over the mouthpiece. Muffled voices. Billy clearly heard the word 'Starbucks'.

Then the obstruction cleared and the female voice said, 'He's unavailable right now.'

Billy thanked her without wasting any more words and cut the line. Great to know that John Vega was making the matter of British security his top priority.

Enjoy your latte.

At the taxi, Mandeep used the radio to tell his control that the policewoman was ready and waiting.

'Officer,' said control. 'She's a police *officer*.'

'Sorry,' said Mandeep to Billy, passing her the handset. She told him it was fine.

The taxi driver took another stroll so Billy lowered herself onto his seat to wait for Dom to come on the line. The humbug jar lay empty on the passenger seat. The boy, Jack, seemed to be asleep in the back. Fluttering eyelids gave him away.

'Alright back there?' Billy raised her voice to be heard through the barrier that divided front and back.

'When can I see my mum?'

'We'll get you home soon.'

'That's what Mandeep says. He's in a mood.'

'Maybe you should let him win a humbug?'

Jack gave a put-upon sigh but agreed.

The radio screeched and she raised the handset to her mouth. 'Hello?'

'I have Superintendent Day in the station for Sergeant Kidd in the vehicle,' said control, sounding very pleased about it. 'Now, I know you officers need your privacy so I'm going to step out of the room for a few minutes and let you speak confidentially.'

'Roger that. Superintendent Dominic Day, reporting for duty.' His spick-and-span tone made Billy smile. *Cheeky bugger.* 'I do appreciate your discretion at this time of heightened national security, madam, over.'

'You're welcome, over,' said control.

'Dom?' said Billy.

'Billy? Over.'

'Pack it in.'

He dropped his voice. 'How you diddling?' His dad's phrase. He'd had a nice dad, but one who dealt with human emotions by reducing them to concepts such as *diddling*. Even at the funeral for Billy's mother when she was sixteen, delivered with a downward inflection: *How you diddling, love?* Dom's voice held the same concern now.

'I'm diddling middling. I arrested a bloke on a drug offence.'

'Nice.' Dom sounded gratifyingly surprised. 'Was that just . . . a random collar?'

'Long story. Unconnected to Dead Driver. What have you got?'

Dom went into a long explanation about how difficult it had been to get information out of Airbnb, but eventually explained that the address given to the car rental firm was a real property, it was really in Howle Green, and it had really been rented in the name of Chad McClusky, paid by a real American-registered credit card. Same card as the one in his wallet. The name might be fake but the payment was real enough.

'Can we get onto the bank?' Billy asked.

'Already got a request in,' he said.

'Can we get a local copper out to the Airbnb?'

'Already got a request in. But literally every pair of police boots in the region is in the city. It's all hands on deck right now.'

'I feel like this Airbnb is relevant,' said Billy. 'Could you get there?'

'We were just discussing it. In this traffic it's going to take a couple of hours minimum, even with the blues and twos.'

'If the traffic clears here any time soon, I'll get there quicker,' conceded Billy.

'Howle Green is literally five minutes from you. And, chances are, it's just an Airbnb he hasn't visited yet and, indeed, never will. A crash pad.'

Billy considered a moment. 'Why Howle Green, though? If Dead Driver wanted a crash pad, why not stay near the airport? That would be more anonymous. Anyway, it doesn't help with finding his identity. What made you think he's not really Chad McClusky?'

'Nothing online under that name. Not with that spelling. The identity is a ghost. So I suspected it must be fake.'

'Is that why you wanted me to call in?'

'Something else. I dug out the file on your jailbird too. The nurse. Florence Nightmare.'

'What about her?'

'She went down for assisted suicide. A mercy killing. Her friend had a degenerative disease. Your nurse got the full tariff, fourteen years, plus a couple of incidents in prison meant she was denied parole, so she served twelve. Sentence was harsh, if you ask me. Hope that judge never needs a one-way ticket to Switzerland.'

Billy took a moment for the information to settle. Pat was a killer. But assisting a sick friend to die is a far cry from murder on a motorway.

The depressing news leeched her energy. Billy wrapped up the call with Dom and trudged towards the grid dragging her jet lag. She should have been home by now. Writing a letter to accept retirement. She'd given her life to her career. She believed in justice. But sometimes it felt too much like bad faith. It wavered. If she was going to avoid writing that letter she would need a sign of her calling.

The brake lights on the back of the stolen SUV flashed red. On-off. Just the once.

Billy stopped in her tracks.

Alright, then, that'll do.

She resumed walking.

There were two heads inside the SUV. The one in the driver's seat was tall and caramel – Charlotte McVie. The other was hidden by the headrest but unmistakably Pat Mackey, the nurse, Florence Nightmare. She could hardly arrest people for visiting each other's cars. But she wandered over and gestured for the younger woman to slide down her window.

'Just checking up on you two,' Billy asked. 'Both alright?'

'Yeah.' Charlotte frowned at the concern. 'How long do you think I can run the engine with the aircon switched on before I run out of petrol?'

'You should be alright. Sun will go down soon.'

The two women resumed staring at Pat's phone. She held it up so they could both see the Wordle App. They'd put in STUCK and got one green, then SNARL, and got another hit. SHAME, thought Billy, but they went with SHADY, and that proved to be correct. Shady. Would the real Slim Shady please stand up? He must be here somewhere, right under her nose.

Her gaze travelled all around the grid, making a big circle to . . .

. . . the space behind the black sedan.

Which had been briefly occupied by a motorbike, according to Carl.

Billy had heard the revving engine of the motorbike herself. Otherwise, she might wonder if he'd made it up. A distraction. After all, there was the strange matter of the burner phone that he'd flung onto the hard shoulder and then picked up again.

Billy went over to his vehicle, the estate agent's bulky Range Rover.

He opened the window at her approach.

'You said this motorbike hared off?' Billy said.

'I was on the phone, so I didn't actually see—'

'I don't understand why nobody saw this motorbike go down the hard shoulder.' Billy turned to face the way the traffic should be flowing as she said this. A few metres away across the carriageway, Daisy stopped on her way to her boyfriend's camper van. She'd ditched the straw hat but gained a pair of joggers and a thin shawl that covered her shoulders. The mermaid dress had given her an otherworldly air of purity – tarnished though it was – but now she seemed a little burdened by the gravity of reality.

'Ah, sorry, officer?' Daisy said.

Billy hoped it wasn't more procedural questions.

The girl explained that she'd seen the motorbike too.

'He zoomed right past me.' She pointed at the gap between her Micra and the central reservation. 'Nearly took my mirror off. I couldn't believe it.'

Billy recalled how Daisy – barefoot and sequinned – had been the first voice she heard after they stopped. She'd shouted something like, 'Did you see that nutter?'

'Do you think he did it, the motorbike guy?' Daisy asked.

Billy didn't answer because she didn't know.

The easy route for a motorbike that had stopped in the inside lane would have been to turn its front wheel sharply left and head down the hard shoulder past the black sedan. But instead, it turned right and weaved across the entire carriageway to find a route out of the jam past Daisy.

Why?

'Did you see the bike too?' Billy asked Olly, who had emerged from his camper van in response to his girlfriend's voice. The motorbike must have passed between his vehicle and the Micra.

'No, I was, ah . . .' He made a complicated motion with his thumbs that Billy took to understand as sending a text.

'Aubergine?' Billy said to him.

Olly went pink.

Oh God, it was probably worse. The actual veg box.

'It was that text I showed you,' said Daisy. 'With the rocket.'

Olly shook his hair. 'I heard it come past, so yah, the bike drove in front of me and past her.'

'Hell of an engine on it,' confirmed Carl.

'Made the whole van shake,' agreed Olly.

Billy looked down the carriageway. The SUV belonging to Charlotte – or more accurately, her husband – was so wide, it narrowed the gap to the left and right. It would have been tight for a motorbike to get through, that was true. But, still, the easiest route would have been the hard shoulder. But it didn't go that way.

Why, Billy?

The easy option must have been unavailable.

Why again, Billy?

The hard shoulder was inaccessible, impassable – blocked?

By what, Billy?

'When you got out to pick up the burner phone,' she asked the estate agent, 'did you see anything on the hard shoulder?'

'Nah, I grabbed it and got back in the car. I thought we might move any second.'

'The road was closed by then.'

'I didn't know that.'

'There were red crosses on the gantry.'

'Look, bab, I can only tell you the truth.'

Billy's phone rang, letting him off the hook. The screen said No Caller ID.

Local lines were still blocked. It must be Washington.

Chapter Twenty-Five

7.28 p.m.

'Am I speaking with Sergeant Billy the Kidd?' John Vega sounded pleased with himself.

Billy said hi, and bit back a line about whether he'd had a nice time at Starbucks.

'How you holding up?' he asked.

This small kindness prompted a dam of emotion to form in her chest and she felt like an absolute arse for begrudging the man a coffee.

'I'm alright, John. I'm alright. What have you got?'

'So we found our guy on the security cameras, moving through the airport, nothing out of the ordinary in his manner.'

'Did he arrive in a taxi? Can we trace him?'

'Yeah, he came in a regular cab, but you don't need that. Just listen up. He used a passport in the name of Chad McClusky to pass through security and, of course, that was the name on his flight ticket so no red flags there. I called my counterpart in your local airport too, and that's the name he used to come through your security.'

'Thanks, our local phones are still down so I hadn't been able to make that call yet.'

'And I called the CIA, but they gave me nothing. Which makes me wonder. Anyhow, your guy's smooth through the airport. Not too early, not too late; doesn't need to hang around, but also not in a rush. I watch him move through the whole air-port. No luggage to check in, got his boarding pass printed out, goes straight through security, got his toothpaste ready in a little baggy. Again, no red flags. Takes a pee – in and out of the bath-room in less than a minute, just enough time to rinse his hands.'

Billy resisted saying, *Lucky bastard*.

'Like I said to you earlier, he avoids the facial recognition, but plenty of them do. Our guy's hand luggage is a carry-on, small, nothing that needs to be inspected by airline staff.'

Billy described the case she'd found in the footwell of the sedan. John Vega confirmed that it matched the one on the CCTV, and she let him go on with his story.

'I'm thinking our guy's visiting England for five days, and car-rying no more than a change of underwear, but each to their own. Some business types travel real light, saves them time. But, any-way, our guy makes a mistake next. *Chad McClusky* pays a visit to Starbucks. So I thought, okay, this is my chance. I get his credit card, right? Get you the bank details, you trace him that way.'

'Brilliant, we can do that.'

'Except, no. No credit card under that name at Starbucks. So I double check the time. I double check the security cam-eras. I see him pay with a credit card – he does the tap and go. So the lady in Starbucks double checks the cash register. We got a credit card payment going through at the exact time Chad McClusky paid for his latte. Except it wasn't paid by Chad McClusky. Do you want to know the name on the credit card that paid for his coffee?'

'I do,' said Billy.

John Vega paused for effect. Long enough for Billy to add two and two and make four.

'That name is—'

One of the oddly shaped pieces of this investigation slotted into place.

'Lincoln Quick,' Billy said.

'You know who that is?' John Vega sounded surprised. And a bit deflated.

'He was pretty big news over here.'

'Well, I guess he would be. Nasty business. I had to look him up.'

'He's kind of a household name in England. For all the wrong reasons.'

'That was some D. B. Cooper conspiracy theory shit.'

'No, John, it was much simpler than that. The guy killed an innocent man and went on the run. The only conspiracy was when his government helped him evade justice.'

Billy should have known the Howle Green connection was too much of a coincidence. Occam's Razor. The simplest explanation.

It wasn't the pub landlord who was the link to Dead Driver, it was RAF Howle.

An American spook had killed a man in Howle Green while riding a bicycle.

A bicycle spoke had killed an American man on his way to Howle Green.

The murder weapon was symbolic.

Billy crossed the carriageway to the sedan.

And there was Lincoln Quick. Dead on an English road, just like the man he had killed. Only this time, Quick hadn't been fast enough to run from justice.

Chapter Twenty-Six

7.31 p.m.

The only connection to Howle Green was the silver fox, Nigel Heathcote. And he'd confessed that he didn't like Americans coming in his pub. Didn't mean he wanted to kill one. But he might know who did.

He was sitting inside his car, she could see the silhouette of his head, the flat top of his smooth quiff. In general, he was very smooth, the pub landlord from Howle Green.

She reached the BMW in the outside lane. Leaning down, she waited while the passenger window slid open. Inside, Heathcote was listening to jazz. The kind that sounded like it was playing backwards.

'Do you like that music?' she asked.

He smiled but kept his eyes on the road ahead. 'I wouldn't play it in the pub.'

Some kind of wind instrument hit an extremely high note.

'It's not exactly pleasant,' said Billy.

His smile widened. He had Hollywood-white teeth. 'It's not supposed to be. It's a rejection of pleasant. After a while, you get used to the dissonance.'

'What does that mean?' Billy asked. 'It doesn't have a tune?'

'Dissonance? Yeah, it rejects our expectation of melody. I find it quite relaxing when I'm stressed. It's unpredictable, so you have to give in to it.' He shifted in his seat so she could see both his eyes. 'You know how some people are into submission?'

This conversation had taken an unexpected detour.

Billy made a non-committal noise.

'Well, this is similar. You . . . submit. Only it's less hassle. Less mess.' He grinned, pleased with the size of his own metaphor. She wondered how many tipsy women fell for that wolfish smile in his pub at the end of the night. Personally, she thought he was a bit of a tit.

He rambled on about jazz, but she tuned him out while her mind turned over that word he had used: 'submission'. Compliance. Deference. Following orders. Lincoln Quick had been murdered right here, but this thing must have started before he reached the motorway. The killer could have had contact with the victim at the airport. It was the obvious place for an attacker to have entered the sedan. And what if the killer had help? Heathcote couldn't have wielded the bike spoke during the actual attack inside the car – he was too far away, sitting in the outside lane – but what if he'd been at the airport earlier giving orders? To someone more submissive.

'So why are you stressed today, then?' Billy asked him.

'Looks like I won't be home in time to open up tonight. Friday night is busy, big loss of takings if I'm closed. What about you, Sergeant? You look worried too.'

She wasn't about to roll over and show him her soft flesh.

'Where have you been today, Mr Heathcote?'

He turned down the wind instrument until it sounded like a squashed insect in its death throes.

'Why do you ask?' he said.

'I'm trying to get a sense of everyone's movements today.'

'On what jurisdiction?'

'A police investigation.'

'I thought you said you were retired?'

'Not yet, I'm not.'

Heathcote sniffed. 'I spent the day at the wholesaler and then at a long lunch with a friend of thirty years. The wholesaler, friend and restaurant will all confirm that I have been accounted for since 11 a.m. this morning. When did our friend die?'

'Two and a half hours ago.'

'Two and a half hours ago, I was sitting right here in the fast lane.'

'Were you at the airport earlier?'

'Nowhere near it.'

'Can you open your boot, sir?'

'Why?'

'Just following procedure.'

There was a clunk and the Beamer's boot lid rose. Billy went to check inside, but there was nothing in the back of the BMW except wholesale-sized boxes of salty bar snacks.

As she stood there, the screen of Billy's phone lit up and it performed its own scat jazz medley of beeps and buzzes. Notification after notification flashed and disappeared. Two hours of WhatsApp messages coming through in an overlapping jumble.

The phone signal was back.

She scanned through and there was nothing new except what she'd already heard from Dom on the radio. Billy excused herself a moment and googled the name 'Lincoln Quick'. No time to waste, in case the signal went again. She found no images of his face – after all, the man was a spy – but there was one long-lens paparazzi photo of him sitting in the sunshine outside a branch of Starbucks. It had been taken when the furore was at its height last year.

She recognised Dead Driver. Wide lips, good bone structure, a distinctive flattened diamond of bone along the top of his nose. Thick, wavy hair. It was him.

The caption said the long-lens image had been taken in Virginia, near CIA headquarters. Figured. The Americans described him as a 'data analyst' but everyone knew that meant 'spook'. Spooks were covered by diplomatic immunity; that was how Lincoln Quick had been able to get away so, well, quick. The headlines she found on Google reminded Billy of the public outrage over the case.

Lincoln Quick had been riding a £5,000 racing bike. All the newspapers were very specific about the price of the bike as though its value was proportionate to his guilt. He often cycled too fast through the sleepy village of Howle. Several articles used the patronising phrase 'sleepy', suggesting that the people of Howle (Howlers?) wandered around rubbing their eyes. But one evening about nine months ago, Quick had raced the bike through a red traffic light and hit a man who stepped out onto the crossing.

The victim was Thomas Kingston. Husband and father. He took six hours to die of internal injuries. By that time, Lincoln Quick had fled the scene, riding the bike back to RAF Howle, where he was bundled out of the country on an American military flight. And then protected by diplomatic immunity. So while the victim's grieving brother led an emotional campaign to have Lincoln Quick extradited to the UK to answer police questions, his killer was free to enjoy a Starbucks in the Virginia sunshine.

None of this told her who had taken justice into their own hands with a bike spoke. Without figuring out that part, she was no closer to leaving this purgatory and getting home where she could hide in her own bed under ten togs of self-pity.

But at least now she had a working phone to check out Heathcote's alibi.

After a few calls to find an out-of-hours number, she got through to the manager of the wholesaler, who confirmed that Heathcote had been there for a good hour that morning buying the bar snacks that Billy had seen in his boot. The manager regaled Billy with his opinion of Heathcote's long-standing reputation as a 'good bloke', describing a pillar of the community who hosted charity events.

Billy didn't point out that, after Jimmy Savile, charity was no longer considered a watertight alibi.

Instead, she thanked him and contacted the friend of thirty years who answered her call on the second ring – obviously forewarned.

Name of Tubby Eden.

Sounded like a plus-sized porn star.

In reality, Mr Eden was a golf club type. Probably drove a Jag. He said, yes, Heathcote had been at the Wharf Tavern on the canal side. Table booked for 1.30 p.m., best seat in the house, they were regulars. Mr Heathcote consumed a beef Wellington and one glass of Malbec – then switched to soft drinks on account of driving. Paragon of virtue. Both men left the restaurant around 4.30 p.m. Again, Eden asserted that whatever Sergeant Kidd was investigating, she must be barking up the wrong tree.

'You understand, don't you, that "barking up the wrong tree" means you are calling me a dog?' Billy said.

Tubby Eden conceded that it was a poor choice of words and quickly rang off, sensing that he wasn't doing his boy Heathcote any favours.

Finally, Billy called the Wharf Tavern. The manager found a credit card payment in Heathcote's name that had gone through at 4.27 p.m. Lincoln Quick had been killed a few minutes before 5 p.m., so Heathcote would not have had time to do anything nefarious between leaving the Wharf and getting stuck in the traffic jam. Such as taking a diversion to the airport and some-how helping to set up a murder. So Heathcote wasn't involved. But now she felt sure that the killer must have entered the vehicle before it reached the motorway. She texted Dom, asking him to trace the sedan between the airport and here, in case it had stopped en route to let anyone get in. Just as she pressed send, her phone rang with his name on the screen.

'Phone lines are back,' he said. 'I heard from John Vega.'

'I spoke to him too. The dead driver is Lincoln Quick.' Billy had been looking forward to giving Dom that news. Pathetic as it was, she needed a pat on the head.

'Inspired idea to contact security in Washington, Billy. This changes things.'

'This murder is obviously connected to the events in Howle Green. I'm thinking about Lincoln Quick's victim, Thomas Kingston.'

'I was looking over that file too. I'm going to call contacts of Thomas Kingston, starting with the family. Establish their whereabouts this afternoon.'

'His brother led a campaign to extradite Lincoln Quick. Can't recall his name, but I remember seeing a picture of him at the White House.'

'I'll start there.'

'Thanks, Dom.' She heard the distinctive sound of the nick in the background, the hubbub of a Friday late shift. A senior officer like Superintendent Dominic Day should be long gone. But from what he'd said earlier about his wife, he didn't have much to go to. Still, he could be in the pub or out on a date. Instead, he was in the station fielding Billy's calls. Maybe he'd stayed late because he cared. Or maybe it was 'the thrill of the case' as they used to call it.

'I got your text about the movements of the sedan. We have CCTV of it leaving the airport parking at 4.46 p.m. – driving time to the motorway suggests he went straight there without stopping.'

'So the killer must have got in the car at the airport?'

'If he was in the car,' said Dom.

'Must have been, I can't work out any other way they could have killed him.'

She heard his breath catch on his tongue as he thought about what to say next.

'So "Chad McClusky" was a fake identity that allowed Lincoln Quick to slip into the country undetected.'

'But why come back to England at all?' Billy asked. 'That's what's bugging me.'

'Unfinished business?'

'Must be pressing business, because he knows – knew – that any-one who spotted him in Howle Green would call him in. There's a warrant out for his arrest.' Billy glanced over to Heathcote, who was standing beside his vehicle again, arms folded like a nightclub bouncer. 'Feelings run pretty high in that village.'

'I should have gone when you asked me to,' said Dom.

'To the Airbnb? You'd only be stuck in traffic like the rest of us.'

Dom clucked his tongue. 'Why didn't he fly in on an American military plane?'

'Because the Americans didn't know he was planning to come? Must be private business?'

'Bloody risky, if someone found out.'

'Someone did find out. They've stuck a bike spoke in the back of his neck.'

Dom made a low whistle.

'There's this weird van on the other side of the motorway,' Billy said.

'Going in the other direction? What's weird about it?'

'Blacked out windows. Hybrid Welfare Unit, it says on the side.'

'What does that mean?'

'Don't know. And the driver is a pit bull. Wouldn't open up. Knew his rights and all that. I don't even have my warrant card on me. He got very sweaty. Threatening.'

'Physically?'

'No, not physically.' Even though she'd feltted aggression coming off him like stink.

'Got the number plate?' he asked.

Billy found the note on her phone and read it to him.

She heard his keyboard clack and her mind flitted to thoughts of his fingers.

Ffs, Billy.

'Fruit farm,' said Dom.

'What?'

'The van is registered to a fruit farm. No red flags, not even a parking ticket.'

She snorted. 'This van has got strong *X-Files* vibes, and we know Lincoln Quick is a spook, so I want to see inside it more than ever. Can you get me a warrant?'

'Tomorrow, maybe. You need to take a look sooner than that.'

Billy sucked in a breath. 'The van driver knows about the scooter thing. He googled me. And he threatened to tell everyone.' Her insides hardened.

'So what?'

'So then I'll have zero chance of solving this case. It'll be dark soon, people are tired and hungry, and too bloody hot. It could turn nasty here if they don't trust me. Even I don't trust me. I might as well retire for all the use I am.'

'People are capable of forgiveness, Billy. Maybe start by forgiving yourself and go from there?'

'What if he tells everyone the story?'

'Then tell them a different fucking story. You can't let one bad day overshadow thirty good years. You're better than that and you know it. Otherwise, you wouldn't have come back from Australia. You don't need me to tell you that, and you don't need a motorway full of strangers to tell you either. You're bloody gagging for your job back, and I'm a bit fed up of you being coy about it. Get back in the saddle, Billy the Kidd. Start with this tosser's van.'

'Jeez, Dom.' Billy rubbed her chest. But he was right. 'Fine, I'll talk to the driver. But I've got no grounds to arrest him so I don't know how I'm going to get him to open the van.'

'Sweet talk, Bill. You can be difficult to resist.'

Chapter Twenty-Seven

7.41 p.m.

It wasn't the most elegant way to approach a suspect, but Billy had little choice except to flop herself over the central reservation once again. She landed on her feet and wiped the muck off her front as best she could. It was mostly grease so it just moved around.

Inside the cab of the Hybrid Welfare Unit, the driver was still listening to his podcast. He moved his head side to side as she approached, cricking his neck in anticipation.

She walked right past his door without looking at him, then made a tight circle around the back of the vehicle. She stopped behind the rear doors.

'Can I help you with something?' The guy's voice was low and close. He'd got out and moved silently to appear around the high side of the vehicle. He was solid with short legs. Like a punchbag wearing a T-shirt. The honed physique of someone who wanted to look hard as well as be hard. And yet Dom said this van was registered to a fruit farm? She didn't recall the Man from Del Monte looking this ripped.

'You solved it yet?' Billy asked him.

'What?' Hybrid Van Man put his thumbs through his trouser loops and pulled his jeans up higher on his non-existent waist.

'The true crime podcast. I think I got it about episode four. Won't give you any spoilers, though.'

'What you doing back here?' he asked.

'I'm just standing here.'

'I told you to stay away from the van.'

'I'm investigating a murder. I would really appreciate it if you would open these doors so I can quickly check—' Billy was interrupted by shouts from further up the carriageway. More than shouts, it was like a chant, rhythmic and guttural. She knew at once that it involved the three Neanderthals who'd been heading the flat football.

Then there was a muffled scream.

Distinctly, a woman's scream.

'You'd better check that out, officer,' he said.

Billy held his eye for a moment, then turned on her heel and walked round the far side of the van in the direction of the fracas.

Along the carriageway, the direction the traffic should be flowing, there was movement. Three pinheads surrounded an estate car in the middle lane.

'There's no need for that, no need.' A shaky voice came from an older man standing in the inside lane beside his own car.

'What you going to do about it?' The younger voice had a sharper edge.

'She's a woman!' the older guy shouted, voice breaking.

'Nothing gets past you, does it? David fucking Attenborough.'

Now Billy could see there was a woman inside the estate car. She must be the one who screamed, her voice muffled because her windows and doors were closed.

Billy had known right away that those three lads had the capacity to go feral. She could smell the pheromones along with their Lynx Africa.

'What's going on here?' Billy asked.

'Is that the copper?' one of the lads muttered. He'd decided to grace the motorway with an unrestricted view of his chicken-skin chest. Fresh sunburn made it look like he was wearing a bib made of ham.

'We ain't done nothing wrong, love,' said his mate, a much taller British Asian lad who sported the substantial beard of a Victorian cricketer that he must surely be ruing in this heat.

'These youths are harassing this woman,' said David Attenborough from the inside lane.

'Sexist,' said the third one, a white lad with a half-hearted mullet who was standing right beside the woman's window. His matching T-shirt and shorts were entirely black on one side and entire white on the other, like some kind of Roadman take on Pierrot the clown. 'Woman can speak for herself, you know?'

He bent over and pressed his face against the window, inches from the woman through the glass.

'Speak,' he shouted. 'Speak!'

The woman cowered away from him. He rapped with a knuckle, like a kid getting the attention of a goldfish in a bowl.

'That's enough,' Billy said. 'Step away.'

He stood upright and adjusted his neckline before moving.

'Fucking bitch won't share,' he said. 'Sharing is caring.' His two mates lowed in agreement.

The woman inside the car was holding the steering wheel, but she looked up as Billy drew alongside. She was middle-aged, with a youthful mop of curly hair and big eyes, made lustrous with tears.

'What is going on here?' Billy asked.

David Attenborough explained that the three lads had been pestering the woman. When he'd stepped in to defend her, they turned on him too.

'Are you alright?' Billy shouted through the window. The woman obviously wasn't alright, and she wasn't going to open her door either. She nodded uncertainly. 'Have they touched you?'

'I ain't touching her,' said the one in the black-and-white T-shirt. 'Shrivelled up old—'

'Shut up,' said Billy with enough force that he did. She pointed at his matching top and bottoms. 'Why the pyjamas? D'your mum order you that off Boden?'

He smacked his lips while his mates winced.

'This is Moncler, lady.'

'Rip-off Moncler,' giggled Chicken Chest.

'I got it off the Outnet!' Pyjamas turned on his mate in protest.

'Out the market more like,' muttered Beard.

With the gang's unity disrupted, Billy resumed. 'Why are you hassling this woman?'

Beard stepped into the role of spokesman. 'Because she's got food. A boot full. Fancy wedding food. Canapés and that. We're so hungry and this is an emergency, so she should share.'

David Attenborough spoke up again. 'She has explained that the food is chilled and prepared for an event, a wedding tomorrow. Surely we can all manage for a few hours.'

Pyjamas strolled past Billy towards the older man, the dark cape of his shadow curving across the carriageway. 'I ain't chilled. I'm fucking hangry.' He delivered the last word at the same time as slamming a flat hand into the older man's chest who staggered backwards against his own vehicle.

'Enough!' Billy got between them. She didn't need anyone having a cardiac. 'You three, get back in your car and stay there.'

'We want something to eat.' Pyjamas stood his ground.

'Yeah, I'm starving too,' said Beard. 'I don't often lose my rag, but I'm feeling it.'

Chicken Chest cackled. God, Billy hated the ones who stand in the background and cackle. Somehow they're worse than the ones who do stuff. She didn't even have the satisfaction of sending the cacklers down.

'Sit in your vehicle and I will find you some food.'

You fucking toddlers.

All three moved into the space beyond the woman's bonnet.

'I'm getting you nothing until you're inside your car.'

The three got into a nice Mercedes parked in front. Chicken Chest in the driver's seat. How does a little numpty like that own a car like this? Injustice never failed to exasperate Billy.

There was a clunk as the woman's door unlocked. She got out onto the carriageway. No more than five feet tall and slender with it, she nonetheless kept her chin up. 'If they stay inside their car,' she said loudly, 'I will find something they can eat. But it will be cold. Everything is on ice.'

The woman opened the huge boot of her estate and moved around polystyrene boxes.

Who was it who said that society was only three meals away from anarchy? Or maybe it was nine meals from chaos? Either way, it was about food; the Covid pandemic had proved that point. A few empty supermarket shelves and people were stockpiling pasta and hoarding loo roll and eyeing each other's pets over the garden fence, wondering if we'd get bolognese for a family of four out of next door's guinea pig. The motorway had got out of hand and they hadn't even missed dinner yet. If it went on much longer, that sweaty pony would start looking tasty.

The woman pulled out the smallest size of polystyrene box. She opened it and lifted an ice pack to reveal trays of tiny finger food.

'I didn't want them rifling through it like chimpanzees at a garden party. My whole business is in the back of this car.' Finger Food flicked her curls out of her face as she held out the box. 'You can give it to them, I'm not. Wagyu sliders with truffle. Bloody wasted on them, but it might shut them up.'

'I know what truffles is!' Pyjama's voice sang out from the car in front.

Billy went and passed the posh burgers through the open window.

'We just asked her to share, that's all,' said Beard. 'She has all the food and we got nothing.'

'I have seen one of you hit a man,' said Billy. 'That's assault. If I see you get out of this car for so much as a pee, I will arrest you. You can go in the damn polystyrene box and hold onto it.'

Not a bad idea actually.

Their window slid shut. Only then did Billy realise that her hands were shaking. This situation could break down in a moment. And it would be getting dark soon. Everything got worse at night. The wolves came out of people in the dark.

She checked on David Attenborough – real name Jeff – who had recovered and was sitting in his vehicle enjoying a much larger polystyrene box. Order was restored, for now at least.

Once again, Billy approached the Hybrid Welfare Unit.

The driver was back in his cab but got out as she arrived. He stood with one hand holding the top of the door, the other hand in a fist on his hip.

'Can I look in the back of this van?' said Billy. 'Please.'

'If you tell me what to do one more time' – he came up so close she could smell mint on his breath – 'I will tell them how you killed that kid on the scooter.'

Chapter Twenty-Eight

7.46 p.m.

I didn't kill anyone.

How many times had she heard those words? Normally issuing from the mouth of some boy racer who had crashed a car with his mates inside or some party girl who had taken her baby to hospital with neglect written on its little body.

I didn't kill anyone.

What they meant was: I didn't set out to kill anyone.

But that's different.

It was a child's response. Someone who hasn't yet learned the hard way about responsibility.

Billy knew all this, but she found herself saying it aloud.

'I didn't kill anyone.'

'What about that lad, then?' said Van Man. 'What about him?'

Billy turned to see everyone standing along the barrier on the other side of the carriageway. The full cast of misfits. Daisy and Olly. Pat and Charlotte. Even the mother and daughter – Kerry and Hyacinth – had joined the gang. Their collective gaze felt hotter than the freakish weather. Billy wanted to protest that she'd been cleared of misconduct, but it wouldn't make any difference. She couldn't deny that the scooter thing had been her fault. Even though she'd left the job, left the country – run away, some might say – she couldn't leave it behind. It had followed her to Australia. Cut through her self-belief until her confidence collapsed beneath the weight of it like a bad knee. It was the scooter thing, not the bloody menopause, that had left her in this state.

'Is it true?' Heathcote asked.

'She didn't kill no one,' said Pat. 'That's the end of it.'

'You told me earlier that you'd googled her?' he said to the nurse.

Oh, she did, did she?

Et tu, Foul Mouth.

'So what did she do?' Heathcote pressed.

'I was lying,' said Pat. 'I never googled her. I said it to shut you up when you kept on with all that "Can we trust her?" rubbish.' Pat tugged at the end of her nose so that Billy couldn't work out if she was lying or lying about lying. 'Keep up, boy, we didn't even have signal when I told you that.'

But Heathcote was right. Policing is a question of trust. And the only way she could win back trust – the most important trust there is; the trust she had in herself – was by doing her job. Because what was the alternative? Retire? A few hours ago, that was her plan. Go home. Give up. But this case only went to show that Billy couldn't even do that – she couldn't let it go.

So, oddly, Dead Driver had given her a new lease of life.

Now that she knew he was actually called Lincoln Quick – a killer who had borne no responsibility himself – she felt no guilt about using him as a vehicle to redemption. She'd solve him like the puzzle he was.

Unlike him, though, she would take responsibility.

She waved Van Man aside and stepped onto the footplate of his cab, leaning over the open door, so that everyone could see her.

'Why don't I tell you what happened,' she said. 'And save our friend here the bother.'

They all settled on the fence to listen.

'It's hard to know where to start, to be honest. So I'll just start on that day. It's a typical Friday. Busy. I'm working single crew, on my own, because we're understaffed as usual. And a call comes in about this lad on a scooter—'

That's not the right place to begin.

It had started much earlier. Weeks, months. A whole career, even.

Billy stepped down from Van Man's cab. Standing up there made her feel like she was in the dock. The drivers lined up along the metal barrier were her jury. Heathcote frowned. Daisy was bright-eyed. Kerry and Hyacinth, the skinny girl fingering the crown of her head again. Pat caught her eye and gave a tight nod.

Billy climbed back up on the step.

'The police take trauma more seriously than they used to. If you go for counselling nowadays, it gets categorised into first, worst or most. "First" means the first time you witness something traumatic. So your first fatal road accident, your first dead body and so on. "Worst" means the worst example of any kind of incident. A worst dead body might be a rotter – one that's decomposed – or a child. And "most" means frequency. In some ways, that's the hardest because it creeps up on you. If you have a whole bunch of fairly mundane fatalities one after the other, it kind of builds up to more than the sum of the parts. Death by a thousand cuts, that sort of thing.

'But what can you do? You can't refuse to go out on call, can you? It's your job. So you absorb it. Like toxins. Convince yourself it's normal, you can handle it. But it doesn't get easier just because it happens every day.'

Over the central reservation, Daisy turned tail and left. Olly made to follow her but she shrugged him off. The part of Billy's brain that never seemed to sleep – like a shark – noted the girl's departure.

'I'm making this into a long story. What I'm trying to say is that it had been a tough few days. I had a first and a worst in one go. A family out walking their dog got killed by a drunk driver. She veered off the road and crushed the lot of them against a wall. All dead, except the drunk who only got a black eye from the air bag. And I knew the family. They had an older boy at university and I had to meet him off the train with the FLO – family liaison officer – take the poor kid back to an empty house. There was a lasagne burnt to a crisp in the oven. They'd left it cooking while they popped out to walk the dog.

'It was a week of mosts too. We had this crappy little gang in town, into vandalism and theft and anti-social behaviour, fights, break-ins, bin fires. One of them got expelled and they were celebrating with a rampage. And their antics, night after night, on top of dealing with this poor bereaved kid, just made me . . .'

Billy looked up and down the motorway. Had a moment of total disconnect. None of it seemed real. Like staring up at the stars in the night sky. What did any of it matter? Lincoln Quick was just one person amid hundreds on this road, thousands in this city, millions in this country – who cared how and why he died? Maybe he got what he deserved and that was the end of it?

'You alright, gel?' called out Pat.

Billy licked a salty film off her top lip.

'The counsellor said that when our brains can't cope with that much sadness it diverts the emotional energy into anger. So, yeah, I was sad. And fucking furious. And on the Friday, when the call comes in about one of the halfwits racing around town on an electric scooter – I didn't hesitate. I'm in the car, straight into town. I spot him speeding along the pavement by the market hall. Friday is market day and there are old dears, kids and prams, and he's weaving between them. Someone's bound to get hurt. And I keep thinking about that family walking their dog who never stood a chance and now this halfwit doesn't care for life or limb either, he's just like the drunk woman. He's going to kill someone and he doesn't even slow down.

'So I switch on my sirens. He's off as soon as he hears me. Down the hill towards the church. I'm in pursuit but he speeds up. Those electric scooters can do thirty miles an hour, more. He's got no helmet, nothing. I could have let him go. I know where he lives. I could have waited and collared him later. But I pursued. Overtook him down the hill and pulled onto the pavement to block his path—'

'You make it sound like it's his fault,' interrupted Van Man. 'Someone died and you're blaming him. You're as bad as the drunk woman.'

'Pipe down!' said Pat.

'Who died?' said Heathcote.

Van Man held up his phone. It showed a newspaper headline, but he read it out loud for those who couldn't see – and to make his point.

'OFFICER ON CONDUCT CHARGE AFTER SCOOTER CHASE DEATH'.

'Papers always stick the knife in,' said Pat. 'They don't care if it's true or not.'

'That headline is true,' said Billy. That silenced them. 'I was investigated for misconduct. It was touch and go, but I was cleared. I shouldn't have pursued him. We're not supposed to chase suspects. Pursuits are dangerous. I was wrong to do that.'

'One rule for you, one rule for everyone else.' Van Man appealed to the audience on the other side of the central reservation, but their expressions were unreadable.

'Let her speak,' said Heathcote.

'Well, there's not a lot else to say. Where did I get to?'

'Young lad racing down the hill on a scooter,' said Heathcote.

'A young lad, yeah. Young offender. He'd been charged with a handful of misdemeanours in the past. He's what we used to call a juvenile delinquent. Thinks he's above the law and also, apparently, immortal because he's doing about thirty miles an hour without a helmet. So I overtake and pull the car up onto the pavement to stop him. Block his way before he hits the junction by the church.

'But I don't account for the absolute idiocy of adolescents. He jumps off the pavement into the road, reckons he can skid round the police car or something, some ridiculous stunt he's seen on a computer game. But instead, his front tyre explodes and the scooter flips him into the air, and he puts his head through my back window. I tried to give him CPR, but he died in my hands. And all I could think about was how his mother would be heartbroken, even if he was a little bugger to the rest of us.'

She didn't add that it had been all she could do not to weep over the lad right there in the road. If life had dealt her a different hand, this boy who lived in her town might have been friends with her daughter, Billy might have got to know him as a parent not a police officer. But the *what if* gave way to the *what is* and she'd had to let go again.

Billy got down from the van on rubber knees. The hot tarmac reeked of burst tyres. That confession was worse than giving evidence to the misconduct hearing. At least a panel of her peers understood the pressures of the job. But her permanent move to 'light duties' – desk bound – had been designed to make sure that she understood it too. No amount of prior trauma that week justified her impatience that day.

'And you kept your job after all that?' sneered Heathcote.

Van Man scoffed in agreement.

'You two must lead a blameless life,' muttered Pat.

'I'm surprised you agree with her,' said the landlord. 'Ex-con and a copper? She'd have put you away for what you did.'

'She'd have had no choice,' said Pat. 'They put me away because I killed someone. My partner was dying, and I did it because she begged and I couldn't watch her shrink and suffer. But I was a bloody nurse, I was hardly going to get away with it, was I? Same as her, Sergeant Kidd, she does what she does despite the cost. So we've got more in common than you think. How come you don't understand people, do you live in a pub or a fairyland?'

'You're all idiots!' yelled Daisy. She marched over from her Micra. Her unexpected outburst silenced everyone. 'He's controlling you.' She waved a dismissive hand at Van Man. 'He's undermining a police officer to direct your attention away from a perfectly legitimate request for him to open his van. It's blatantly sexist what you're all doing. You wouldn't second-guess a male officer—'

'Don't play the sex card!' said Van Man.

But Daisy was on a roll. 'Controlling men find a weakness and use it against you. He's controlling us and we're letting him. I for one want to know what's in his van.'

There was a long moment filled only with the rattle of an empty crisp packet blown along the hard shoulder. An unruly gust that presaged the collapse of a too-hot day.

Billy turned to face Van Man.

'I want to know too,' she said.

There was a murmur of agreement from along the central barrier.

Van Man moved both fists onto his hips.

A scuff of footsteps alerted Billy to movement behind her. The older gentleman, David Attenborough, and the woman with the catering business, Finger Food, had drifted along the carriageway, drawn to the tension.

Chicken, Beard and Pyjamas arrived too. These three useless items stood sentry over her right shoulder, like pound shop angels. She seemed to have been appointed leader of some *Mad Max* ragtag army. Not quite how she'd envisioned her atonement going down, but that's how it was playing out.

'Open your spooky van, guy,' said Beard.

'Or we'll do it for you,' said Pyjamas.

The older man chipped in: 'I really do think that would be the quickest way to resolve this situation, sir. If you have nothing to hide, then open up.'

'Why you all listening to her?' Van Man rediscovered his voice.

'Look.' Billy stepped forward a pace and felt the pound shop angels shadow her move. 'I have not laboured this point until now because I don't want to start another panic.' She nodded conspiratorially towards the vehicle containing the man who had panicked earlier and mounted the central reservation like Putin on a horse.

'That pussy,' muttered Chicken.

'But there is a killer among us. A man has been murdered in his vehicle and the assailant is still here.'

'How do you know that?' The older man, voice aghast.

'Because where else has the suspect gone?' Billy opened her palms as though she might magic him up. 'He hasn't climbed over the fence, it's four storeys high' – she pointed towards the sound wall that lined both sides of the carriageway beyond the scraggly trees – 'and if he'd run off down the motorway, someone would have seen him. He must be hiding somewhere.'

'How do you think he got in my van?' asked Van Man. 'He would have had to climb over the central reservation in full view of everyone.'

'There is someone in the back of your van,' said Billy. 'I saw a shadow move in the side window earlier and I want to eliminate it from my enquiries, that's all.'

'It's a dog,' said the Van Man.

'Why won't you open the van, guy?' said Beard. And then he muttered to his pound shop angels, 'He's well suss, this one.'

Van Man fished in his pocket and in one jerky motion, threw a set of keys over three lanes of traffic into the grassy verge beyond.

'Well, that was stupid,' said Pat, who had hopped over the barrier to be on the same side as Billy. She stood at her left shoulder.

'This is lame.' Beard turned on Billy. 'Why don't you open his van? You're a copper. There's a killer in there.'

'I can't forcibly enter a vehicle unless I've got permission or a warrant or I put the owner under arrest.'

'Arrest him, then!'

'I've no grounds. If I force my way inside, then any evidence would be inadmissible.' Billy couldn't help but smile at their incredulous faces. 'Policing is harder than it looks on the telly.'

'Too right, it is!' said Chicken, rubbing his skinny chest.

'She only needs to see inside, mate, open up,' said Pyjamas.

The pound shop angels advanced on Van Man who raised his fists as though to take on all three at once. Billy got between them; she did not want a fight on her hands. Pat got into the fray too, pushing Chicken aside.

There was a scuffling of feet as everyone jostled everyone else, until a loud clunk silenced them.

The older man slid out of the cab. While everyone had been involved in fisticuffs, he must have got into the driver's seat and unlocked the rear doors.

'You can't enter the vehicle illegally, my dear, but I can,' he said. 'Arrest me later if you like.'

'Nice one, Sir David,' said Pyjamas.

Chapter Twenty-Nine

7.52 p.m.

Billy left the driver penned in by Pat and the pound shop angels. She walked around the weird van to the sliding side door that had clicked unlocked but remained shut.

She wasn't sure what might be inside. Had no experience of the CIA beyond reruns of *The X-Files*. She doubted it would be Spooky Mulder hiding in there.

More's the pity.

But Lincoln Quick was an American intelligence agent, who had chosen to return to the UK despite his government sticking its neck out to protect him via diplomatic immunity. His case had sparked an international incident and he'd thrown all that away to come back here.

Why?

Whatever the reason, it was possible the Americans had got wind of his betrayal and taken matters into their own hands.

Was that possible?

Was the bike spoke a message? If so, who for?

Staring at a normal van door, these questions felt slippery. Conspiracy thinking. But instinct had drawn her here. After thirty years in the force, she had a radar for fuckwittery and this dubious vehicle was pinging.

The van rocked slightly and she took a step back. It rocked again. 'Open up!' said Billy.

The door slid open slowly . . . and then very fast.

A blow sent her backwards. The sun burst as her head hit the tarmac. Instinctively, she pulled her legs tight against her chest as dull thuds of pain registered all over her body. People were streaming from the open door of the van, trampling Billy underfoot. She rolled from her fetal position onto her feet. As her vision cleared, figures flitted away into the lowering sun. They could be huge bats.

She lunged forward and grabbed one of the stragglers, catching someone in a rough hold. She clung on, digging into the fabric and flesh of a thin arm. Billy swung the person around, until she could use her superior height and weight to pin the captive against the bonnet of a car.

A woman. Hair flying as she thrashed. Billy braced to hold on, leaning back to avoid receiving a head butt.

'Calm down,' she said. 'Stop fighting.'

The woman froze.

Billy held on a few moments longer. Then moved to adjust her stance. The instant her grip relaxed, the woman's arms wrenched and she burst into action again. She wasn't giving up.

'You're going nowhere,' said Billy. 'I've got you. Just stop.'

Now the woman went limp, flopping forward so that her chest and head lay on the bonnet.

Billy held onto the woman's wrists, panting and sucking air between her teeth.

She realised that the car she was leaning against contained the panicking man from earlier. Calvin Something. His girlfriend made eye contact. Billy felt a twist of guilt because she hadn't checked on them sooner, but then she'd been occupied . . .

'You both alright?' Billy said. The girl gave a single nod.

Calvin Barnes, that was it.

'You're doing well, Mr Barnes, you just stay where you are for now.'

Calvin stared down at his lap. The girl widened her eyes and there was a shared understanding that the couple would stay inside the car while Billy sorted this out.

'What just happened?' Pat appeared around the van.

'Did you see how many there were?' Billy asked.

'They all ran,' said Pat. 'Must have been half a dozen.'

'What about the driver?'

'He legged it n'all.'

Billy gave the woman's hands a gentle tug, to indicate she was speaking to her. 'I'm going to turn you around. I need to ask you some questions, alright?'

The woman said nothing. Billy could see only her profile as she lay with one cheek against Calvin's bonnet. Her hair was wet with sweat, strands lashing her face. The woman stared straight ahead and blinked.

'I'm going to move you to the van, okay?' Billy said.

The woman said nothing.

'Okay, here we go.' Billy half dragged, half lifted the dead weight of the woman to the open side of the van. Its floor hit the back of her knees, folding her legs. The woman landed heavily but safely on her bum.

Now, Billy saw that it was not a CIA van. Unless the CIA had suffered severe budget cuts. Instead of surveillance equipment or whatever high-tech wondery Billy had been expecting, the back of the van was lined with cheap plywood. Wooden benches ran along the sides. There was rubbish all over the floor, mostly wrappers from cheap snacks.

If you imagined a van that might be used in a kidnapping, this was it.

Billy knew at once what was happening here.

'You're not in trouble, love,' she said.

'No English,' said the woman. Her eyes were defiant.

'Where are you working?' Billy tried again.

'Illegals,' said Pat in the background.

'They were held in there like cattle,' said David Attenborough.

'Welfare unit, my large arse,' said Pat.

'What kind of work are you doing?' Billy asked the woman. 'Farming? Nails? Packing?'

She said nothing but glared up at Billy.

'Where are you from?'

Steely gaze.

'I can help you,' said Billy.

The woman's eyes pooled. 'My daughter,' she said.

Billy looked along the carriageway. 'Your daughter went with them?'

'Yes.'

'How old?'

'Sixteen.'

Billy sighed.

'Let me go,' said the woman.

'I can't do that,' said Billy.

'She is alone. Sixteen.'

Billy turned round to Pat. 'Did you see any young girls?'

Pat shrugged.

'What about the others, did they see anything?'

Pat went to have a confab with the angels.

'The driver?' Billy asked. 'What is his name? He said Bob, but . . .'

The woman shrugged.

'If you help me, I can help you.'

'I work.'

'What are you working on?'

'We pick flowers, pack chicken, more flowers.'

All this for a bunch of flowers.

Pat reappeared with the pound shop angels trailing her.

'They saw a couple of 'em,' she said, meaning the illegal workers. 'Says there was mostly girls.'

'My daughter.' The woman's mouth broke into a grotesque smile that quickly collapsed into a wail. She sobbed and then fell silent as suddenly as she started. Carefully, Billy rocked the woman back to look her in the face.

'I need you to stay here and wait for more police to arrive. They can help you find your daughter.'

'You police?' asked the woman.

'Yes,' said Billy. 'I am.'

Billy persuaded her to move into the cab where she could sit comfortably. She locked the doors and left her there. Billy heaved herself back over the central reservation into the grid of cars. The drivers on her side of the carriageway had drawn into a huddle of whispers. They broke apart as Billy approached.

'She'll get out and run,' said Pat.

'If it was my daughter in danger, I'd run after her too,' said Billy.

Pat, reading the unspoken words like tea leaves, said, 'You got a girl yourself?'

'I did, yeah.'

Pat knew to leave it there.

Young Daisy, less attuned, said, 'She could be tugging your heartstrings, this woman?'

'They're broken, love,' said Billy. 'She'd have to be a maestro to get a tune out of me.'

She looked around. They were all here now. Except the estate agent in his fancy car, and Kerry and Hyacinth who were heading back to their minivan; they had each other, didn't need to be standing in the hot sun for a bit of comfort.

But where was Charlotte?

The driver of the stolen vehicle was gone.

No one had seen her since before the confrontation with Van Man.

Billy climbed inside the SUV for a look-see. Again struck by how neat it was, she found nothing but a blocky car key in the centre console. No house key.

Okay, so she'd dumped the stolen vehicle and left the scene to (somehow) go home. She'd been stressed about her kids. But she must realise that with no traffic moving she wouldn't get home for hours. Made no sense. Unless she was running from justice?

Chapter Thirty

7.58 p.m.

The stolen SUV was clean and empty – even the parking ticket had gone along with Charlotte. As Billy got out, her boot kicked something tinny in the footwell. She picked up a canister. A criminal identifier spray.

Standing on the carriageway once again, Billy turned the can in her hand and gave it a shake. It felt quite full.

'Pepper spray?' said Pat. 'Woman I shared a cell with went down for using mace on her rapey boss. Bloody travesty of justice that was.'

'Mace is illegal. This is a dye. You spray it over an attacker and it sticks around for days so police can identify him.'

'What, after he's raped you? Bloody lot of good, that is.'

Billy agreed with a hum.

She'd seen women use some strange devices to feel safer than they actually were – everything from the classic key clenched in a fist, to panic alarms disguised as necklaces, to lockable chastity pants that are supposed to be impossible for a rapist to pull off but would probably aggravate him enough to kill you instead. She knew about dye spray but rarely saw it in action. Why did Charlotte feel the need to arm herself?

'Hello?' The mother and daughter were standing together beside their minivan. It was Hyacinth who had called out. She pointed down the hard shoulder, against the direction of traffic.

'The pretty lady went that way. She was running. I watched her go.' The girl didn't wait for a reply but walked around her bonnet and got in the passenger seat once again.

'Where does Charlotte think she's going?' Billy mused to Pat. 'I saw you two talking, did she say anything?'

'Husband sounds like a right shit.'

'I got that from him reporting the car stolen.'

Billy set off along the hard shoulder after Charlotte. The heat had drained out of the day, which was a relief until she remembered that meant it would be getting dark soon. The Lord giveth and the Lord taketh away. Passing the cordoned-off suspicious car, Gammon still waited at a safe distance on the grass verge. He told her that a woman had come running past about fifteen minutes ago. Billy picked up the pace.

As she jogged, she placed a call to Dom who picked up on the third ring.

'Charlotte McVie,' Billy said without preamble, knowing she wouldn't be able to jog-walk and chat for long. 'The one in the stolen SUV. Tell me about her.'

Dom reeled off vitals – date of birth, maiden name, home address, none of which suggested a link to Lincoln Quick. She was a decade older than him, English and lived a good fifty miles away.

'Shit,' Dom said, making the word last a full two seconds.

'What now?'

'She's got a non-molestation order on her husband. Literally been issued by a court today. Charming Chester McVie has been knocking her about.'

'She had a defensive dye spray in her car. Must have been planning to tag him if he broke the non-mol and approached her.'

'The address of the family home isn't named on the order, so she must be living somewhere else.'

'I knew it was weird, him reporting her for stealing valuables from the house. He must have realised she'd left him. A lot of things make sense now. Even the Rescue Remedy . . .'

'What's Rescue Remedy?' said Dom.

'Herbal sweets that calm you down.'

He grunted. 'Sounds like it's Chester who needs calming down.'

'She's done a runner from the crime scene. Abandoned his car with the key inside. At least we know this one doesn't have a bloody car bomb in it.' Billy's breath was short and she was chasing a woman who was fit, a runner. No sign of her up ahead.

Billy had a thought. 'What about the parking ticket?'

She heard Dom tapping his keyboard. 'The Audi was issued with a notice this lunchtime . . . at a residential address about thirty miles south of where you are now. That's a long way from her home.' More tapping. 'Oh, shit.' This time the word lasted even longer.

'And?'

'The address has got a flag on the system. It's a safe house.'

'What kind?'

'Women's refuge.'

'That's where she got the ticket?' Billy asked.

'Yeah.'

'Oh, Christ. She said the husband was away last night, so today must have been her chance to leave. So she left home, served the husband with the non-mol and used his car to move her stuff to the women's refuge, maybe planned to get the car home before he discovered it missing.'

'Only she got stuck in the traffic jam.'

'No wonder she's in a state. Can you find a mobile number? I need to speak to her. And we need to find out where her kids are.' Billy stopped. She still hadn't reached the end of the high sound barrier. Even if Charlotte reached an exit from the motorway, what good would it do her? She'd see all the traffic backed up and realise it was pointless. She'd probably come back of her own accord.

Billy stopped. She had to let her go. This had nothing to do with Lincoln Quick. And he – or at least his killer – was Billy's priority.

Of course, Charlotte might ask why she was never a priority. Did she have to die too?

Dom came back with a mobile number and Billy dutifully copied it into her phone. Charlotte's words from earlier echoed; *Fucking police*, she had said, *questions after questions and nothing happens.*

And now Billy had to let an abused woman go off alone while she stayed to solve the murder of a man who had himself killed someone innocent.

Where was the justice in that?

Was it even about justice anymore?

Or was it about the needs of the many versus the needs of the few? Someone on this motorway was armed with a weapon they had literally up their sleeve. One person was dead. If anyone had to confront that killer, it should be a police officer. It should be her.

She'd earned their trust, now she had to pay it back.

'Billy?' It was Dom still on the line, his voice so taut it made her picture a telephone cable with all its curls pulled straight. 'Don't panic, but . . .'

'That makes me want to panic.'

'Is everyone keeping a distance from that suspicious vehicle? The one we decided didn't have a car bomb inside?'

THE FOURTH HOUR

Chapter Thirty-One

8.00 p.m.

Billy stood way back from the abandoned vehicle on the grass verge. She tried not to think about the impact of a bomb going off. If it would ricochet off the high barrier behind and effectively blow her up twice. But the cars down on the carriageway were full of petrol, so she didn't want to be among them either.

Swings and roundabouts.

Along the road to her right, the gaggle of drivers who'd been banished from their own vehicles for hours now were shooting her dark looks. She waved a hand for them to stay back.

Down the phone line, Billy heard Dom praise the junior officer who'd found new intelligence on the Police National Computer. Billy tried not to mind that she was the one facing a potential car bomb. She didn't mind really, it was just the urgency of her need to wee. Her nethers hurt. It made her angry and distracted. Angry that she was distracted.

'Dominic?' she said in a tone that would make a dog think twice about eating a bone.

'How far from the car are you, Billy?'

'Far enough. Just tell me, is this car linked to a terrorist cell or not?'

'Keep people away.'

'They are away. They've been away since the last time I said it might blow up, but it didn't blow up and they're not going to believe me forever.' A beat. 'And I told them about the scooter thing.'

Dom tutted.

'What?' said Billy. 'It was better than someone else telling them.'

'You're a copper, they don't have to like you.'

'They have to trust me.'

Dom muttered under his breath, 'You have to trust yourself.'

'It's hard, after the scooter thing—'

'It's got nothing to do with the bloody scooter thing, Bill, and you know it.'

From somewhere, a waft of radio sounded the urgent drums of a news bulletin.

'So what is it, then?'

'You want me to say?'

'Yeah, I want you to tell me how my problems are even worse than standing beside a gridlocked motorway with a murderer and possibly a car bomb.'

'Well, you're not going to like it, but – I think it's the other thing.'

A bubble formed in her throat.

'What other thing?'

'Your daughter. Abigail.'

Billy's voice came out crushed. 'That was twenty years ago. I was young . . .'

'Exactly. I know it was awful, I know—'

'No, you don't know. Your kids are at university, they'll have jobs, marriages, babies of their own one day.' Her eyes strayed to Daisy, who was doing nothing remarkable except standing barefoot beside her own car, clad in the timeless accessory of youth.

Like every young woman, she was oblivious to her loveliness. But she appeared to Billy like a green shoot and a freshly baked bun and a bloody sunrise all rolled into one. The effect hit her with the same gut-punch of grief that she felt every time she met someone around the age her daughter should be now. Any attempt to articulate this sensation to Dom would end in nothing but a puddle of frustration on the tarmac because the feeling was such a jumble of contradictions. Heartache and envy. Sorrow at losing a child. Spite at losing who she should have become. And there was guilt because she should have checked her baby sooner, or laid her down differently, or fed her earlier or later, or cracked a window so it wasn't too warm, or swaddled her tighter, or somehow changed any one of the minute factors that resulted in her baby simply failing to breathe while taking a nap. And there was some indefinable shame at being a victim. And, finally, there was the long choppy wake of a trauma that never truly settled, her anxiety. Because however cleverly a counsellor span the line that amounted to 'it might never happen' she knew better than anyone else that it might, in fact, just fucking happen. It might happen in the middle of the afternoon on a sunny Tuesday when you're planning to wake the baby for a stroll to the park before your night shift starts. So try telling the anxiety to stop ringing its alarm bell, because the anxiety will not be told.

'My point stands,' Dom said. He could never accept that wanting something to be true didn't actually make it true. 'The problem is that you stayed in that bloody town and . . .' he fizzled out and she heard a clunk like he'd hit the desk.

'It's a good town. It's where we grew up.'

Nothing.

'What, Dom?'

'You gave up.'

The bubble hardened into a shot that she fired at him. 'Fuck right off!'

'Come on, you've been grieving non-stop for twenty-odd years. For Abigail. For your marriage. For your career. You were so ambitious, Billy. That's why I followed you into the force, you were Dorothy and I was the scarecrow.'

'What are you talking about now?' All this emotion was making a headache form behind her eyes. 'What does that make the police force, the mighty Oz? All smoke and mirrors?'

'It's a bad metaphor, but it stands. I followed you because you knew exactly what you wanted when I didn't have a clue. And you lost your mojo, for understandable reasons. But the worst thing is, you never went looking for it.'

'I've been a sergeant for twenty years, it's hands-on, it's what I like doing.'

'Is it though?'

'Yes. And anyway, you didn't follow me, you were on a mission of your own. Running after something better, someone better. I was just a prop, a wingman. Like all those years I sat in the passenger seat of the bloody Volvo.'

Suddenly, Billy had to batten down the hatches ahead of an inexplicable urge to cry.

Why now, after all that stuff about babies, am I about to cry over a bloody Volvo?

Dom picked up the baton. 'And that's why I had to go to another force. All those years of driving you around and you never took me seriously enough to—' He stopped and tutted. 'This is not the time . . .'

'You think? What's the upshot of all this, Dominic?' she said. 'The clock is ticking . . .'

'Don't be hard.'

'You know I'm not.'

'The upshot is just . . .' he sighed. 'Trust yourself.'

'Is that it? All that heartache in return for a bloody fortune cookie? Trust yourself? It's not exactly groundbreaking advice. It's not exactly *Lean In*. I don't think you're going to publish those wise words to great acclaim.'

'Okay, then, I mean take a chance, Billy. Take a risk, commit, go all in, stop holding back and hedging your bets and thinking that if you play it safe you can keep your head under the radar and never get hurt. Because you're wasting yourself. Give life a hundred per cent and it'll give you a hundred and ten per cent in return.'

'You sound like a football manager. "A hundred and ten per cent". The maximum is, by definition, a hundred per cent.'

Down the line, Dom cleared his throat but said nothing.

Fine, she'd give him a hundred per cent. But not a per cent more.

'Let's get on with it, shall we, Superintendent? What do you see on the PNC about this abandoned car? Can we *go all in* on that while I'm standing here face to face with a potential car bomb and my head very much on the radar?'

Dom sighed and took a moment to answer. But when he did, he was back in police mode. 'According to the PNC, in 2003 that car was listed as a vehicle of interest following a raid on a property. It had been parked in the street outside. During the raid, the terrorism unit arrested three men related to a lad being held at Guantanamo Bay.'

'Right. Sounds very 2003.'

'Except . . . the vehicle didn't belong to the terrorist cell. But it stayed flagged on the system anyway. It doesn't say why, but there's a code, which if I'm not mistaken . . .' Dom sighed.

'What, Dom?'

'I'll have to call a mate in Intelligence, but I think it must have been a covert vehicle.'

'It belonged to an undercover officer?' Billy stopped. 'Someone infiltrating the cell?'

What was it Daisy had said about the CIA? Forty per cent of their work to prevent terrorist attacks on US soil took place in the UK. Because Brits breed baddies. If there had been a covert operation to infiltrate a terrorist cell in England, then perhaps the Americans had been doing the spying.

And who do we know was a US spy?

Dead Driver.

'It could be a CIA vehicle,' Billy said.

There was a long silence during which Billy heard in one ear the harmony of the wind and in the other ear the disharmony of the nick.

'It makes sense,' she said. 'The CIA could have infiltrated that cell, but they would need our terrorism unit to stage the actual raid.'

'And the CIA operative's car was parked outside when the attack went down.'

'So British police flagged the vehicle—'

'Because they knew it was covert.' Dom clucked his tongue. 'But does that mean it's *still* a CIA car? That raid was twenty years ago.'

'The car's of the right vintage, it's knackered. Who's it registered to?'

Tapping on a keyboard. 'An air conditioning importer.'

'Seriously?' said Billy. 'You don't earn much from making English people colder. Well, except today perhaps.'

Dom gave a cursory huff of laughter. 'Where has this vehicle been for twenty years? And why is it there on the motorway now?'

'Well, we've got a CIA vehicle following an American who also worked for the CIA . . . I'd say the woman who abandoned this car had been tailing Lincoln Quick until the moment he was murdered and then she left the scene. I don't have much of a description of her, though, just a young woman in a headscarf. All they seem to have noticed was the headscarf.'

'Could she have killed him, though?' Dom asked.

'Can't have done,' Billy said. 'Not unless her covert training made her able to turn invisible. Someone would have seen her walking past a long row of cars to reach the sedan and then going back again. It's just not possible.'

'Maybe she was tailing him from the airport to see where he was headed? Maybe his UK trip wasn't as secret as he thought it was? Maybe he wasn't a very good spy. He certainly wasn't a very good cyclist . . .'

Billy hummed in agreement. 'So this woman was following him but when he got bumped off she legged it – she knew the flag on the car would show up eventually in the investigation into his murder, and that would have blown her cover.'

Dom switched to a bad American drawl. 'So she got out of Dodge.'

'Maybe she was assisting the killer? Are we looking for someone else covert?' Billy's gaze slid down the carriageway and came to rest on the slight shape of Daisy. The student who was too interested in police procedure. This case had sent Billy up and down the carriageway more times than a traffic cop. But maybe she should have stuck to her grid. Her first instinct had been that the killer couldn't have gone far.

'Can you check someone out for me?' She told Dom the girl's name and the university she was supposed to attend. 'I want to know if this kid is a student or a suspect.'

Maybe it was his pep talk or the failing light, but something crept out from inside Billy, a nocturnal animal emerging from its burrow; cautious but driven by hunger. She needed to nourish herself by solving this crime.

Chapter Thirty-Two

8.08 p.m.

Billy dialled Charlotte's phone. It rang three times then went dead. Call rejected. If only Billy knew where the woman was headed . . . Presumably, to the children. That gave Billy an idea. She jogged over to Mandeep. He got his control on the taxi radio.

'I don't suppose you made a note of the number when that mother called her babysitter?' Billy asked.

'Yes, Sergeant Kidd, of course I did,' her voice crackled over the radio. 'I log all numbers and addresses, it's my policy. I do love the crime shows . . .' She chattered over an efficient-sounding rustle of paper and then said, 'Have you got a pen?'

Soon, Billy had a number for Charlotte's babysitter. Billy tried it right away. A woman picked up with a hesitant greeting.

Turned out she was called Jan and was looking after the kids, Harry and the twins.

'Has something happened to Charlotte?' Jan asked, her voice hitched in preparation for the worst.

'No, I've just called with a couple of questions. As far as I know, she's fine.'

'She called about half an hour ago, said she's trying to get here but there's no traffic moving. She was very upset. I was only supposed to watch the boys for two hours, I was going to bingo, but I've put them to bed in the spare room. They're no trouble, good boys, which is a blessing.'

'Can I ask where you are? We would like to speak to Charlotte.'

The woman let out a breath that caught in her throat. 'I don't know if I should say. Because of her husband. I don't want to . . . be responsible.'

'You're right to be cautious,' Billy said. 'I assume you know what she's been doing today?'

'She's leaving him. Leaving a great big house too, but you never know from the outside what goes on behind closed doors.'

Billy knew there is stigma about leaving your husband, even a violent husband. The woman had her pride, and who could begrudge her that. Billy gave Jan her own phone number.

'As soon as you speak to Charlotte, could you ask her to phone me? I don't think she trusts the police much, but—'

'Well,' said Jan, making the word last three syllables. 'I'm not surprised. Had a shocking time. They kept saying she needed proof. Well, he wasn't daft, he didn't put her in hospital, always kept it reasonable.'

'Reasonable?'

'I mean not so bad that she needed medical attention. Reasonable isn't the right word. Restrained?'

He restrained himself from beating her too hard. And yet he couldn't restrain himself from beating her at all.

Jan tutted at herself. 'I don't think there is a right word. But the police said they were snowed under with cases like hers. Domestics. Stalkers. Sex pests. It's like burglars, isn't it? I got turned over last year and they don't even come round for a run-of-the-mill burglary nowadays. Only if it's aggravated. Same with Charlotte – he didn't do anything bad enough for them to get involved. She was quite annoyed and her solicitor said the same, you can't rely on the police, and the women's advice line said it too, they won't do nothing to help you, so that was that. She had to sort it out for herself.'

Billy massaged the frown lines at the top of her nose. No wonder Charlotte had told her to fuck off out of her car.

'Did she say why she'd left the motorway?' Billy asked Jan.

'She was triggered.'

'Triggered?'

'Emotionally, you know. It's not the same as overreacting.' Jan spoke as though reciting something she'd had carefully explained a number of times. 'Someone was shouting about a dead lad and it triggered her and she needed to get back to her boys. She wanted to get here before her husband works out where they are. But if he comes here, I'll send him away, I told her that.'

'Do you know him, then?'

Jan laughed, low and mirthless. 'Yes, love, he's my son. We don't speak, I'm afraid. It's broke my heart, but Charlotte sneaks the boys to see me now and then without him knowing. It'll take him a while to think of coming here to look for them.'

'Can you get Charlotte to call me? Please? I want to help her.'

'I don't know if I'll see them again after today.'

'You'll see them again, Jan. She came to you for help, she must trust you.'

The mother-in-law gave a big wet sniff which was the first Billy knew she was crying. There were a few deep breaths while she got her voice under control. 'She's in a state about this parking ticket.'

'On the Audi?'

'It'll lead him to the women's refuge, won't it? When the fine comes through the post, it'll go to Chester because it's his car, and it'll say where the ticket was issued. That was right outside the refuge on the day she left him, so he'll know where to find her. She was trying to pay it on her phone so it never came through the post, but she couldn't because she's not the registered owner of the vehicle or something.'

Billy let her head fall back. Overhead, a slab of cloud was building up, slung low like the suspended ceiling in a grotty old police station.

So that was why Charlotte had been bashing away at her phone. And why she'd been in a tizzy when the signal went down. Her careful escape plan had been ruined by a parking ticket.

'I can make the ticket go away,' said Billy. Here was one helpful thing she could do for the woman. 'Tell her not to worry; I'll get it wiped off the system. The husband won't find out where she's going.'

'You can do that?' asked Jan.

Billy fervently hoped that Superintendent Dominic Day could. She banged out a text message with instructions for Dom to make it happen – or else.

Chapter Thirty-Three

8.16 p.m.

The overcast sky slung low over the motorway turned the colour of an old bruise as the sun started dropping in the sky. Off in the distance, streaky clouds dropped jellyfish tendrils.

'The sky looks like it's melting,' said Daisy.

'Dry storm,' said Pat. 'It's so hot the rain is burning off before it reaches the land.'

'Virga,' said a voice behind Billy. Hyacinth. She'd emerged with her mother to edge into the group standing behind Billy's car in the middle lane. The sinking sun brought them out of their isolation, an instinctive urge to join others for safety. To roost. 'That's what you call it. A Virga.' The young woman's voice was like her body, thin and pinched.

Kerry had a blanket around her shoulders. As the wind got up, the temperature was dropping rapidly. The contrast to the stiflingly hot day made it nippy. The scraggly trees shivered their leaves.

Billy checked her watch. They'd been stuck here over three hours now. People were tired and hungry. They'd soon get chilled, sweat on their bodies evaporating like the virga to lower their core temperature.

She needed to keep her eye on the target.

Howle Green.

The killing of Thomas Kingston.

She left the drivers chatting and sat inside her vehicle with the door ajar.

A Google search brought up thousands of news stories about the Howle Green incident. Too many. She adjusted the search to the actual day that Kingston was killed, before the connection to the US airbase was made public. Now there were very few stories. The death of one man on the road wasn't exactly of national significance. But a couple of local outlets covered it; a pedestrian killed by a cyclist was half-decent clickbait.

She tapped a headline and waited for it to load.

Outside, the group were in dispute. Their voices carried and Billy kept her eye on them in the wing mirror.

'Not sure how we upset Charlotte so much she fled the scene of a crime,' Heathcote was saying. 'Makes you wonder.'

'Makes you wonder what?' Pat challenged.

'What was preying on her mind,' he said mildly. 'To run off like that.'

'She hasn't run off, she's gone to her kids,' said Pat.

'Well, she's not going to get very far.'

Fair point.

On her screen, Billy watched blocks of text appear and – annoyingly – move around too fast to read as pop-ups and images loaded. She dropped her hand to her lap and resolved to wait it out.

'She did seem neurotic,' Olly chipped in. 'Reminded me of my mother.'

'Your mother has just been appointed a judge,' Daisy said. 'She's a total girl-boss. And it's not neurotic to be afraid when you only have experience of police mistreating you. It's like the

African American community in America, or women here after the Sarah Everard murder—'

'I don't see what that's got to do—' the landlord tried to interrupt.

'It's got everything to do with it,' Daisy said, small hands chopping the air like chef's knives. 'It's got to do with trust. With power comes responsibility.'

'Did you learn that at university?' asked Heathcote.

'It's Voltaire via Spider-Man,' she said.

The pair bickered themselves into a stand-off and stood in silence for a moment.

'You two wouldn't last five minutes inside,' muttered Pat. 'You have to learn when to trust someone, when not to. It's not about judging them on everything they've ever done, it's about what they're doing now.'

Amen, sister.

On Billy's phone, the page finally loaded.

LOCAL MAN KILLED IN CYCLIST COLLISION

The headline of the local paper was mild. No sense of the furore that would blow up in subsequent days when it emerged that the cyclist had skipped the country. Billy scanned the piece. The hack had yet to link the accident to the US airbase at all. No mention even that the cyclist had jumped the red light – that detail probably hadn't been released yet. The victim wasn't named, the journalist only confirmed that the man was local to Howle Green. There was a whole paragraph about a controversy over the recently built bypass on which the incident occurred, which local residents complained had divided the village into two. Finally, in the last line, Billy found what she was looking for.

'*Sergeant Gavin Glynn appealed for witnesses . . .*' blah blah blah.

Sergeant Gavin Glynn. First officer on the scene.

Billy WhatsApped the name to Dom with a request for a mobile number. But when the message showed only one tick, she took matters into her own hands. A few more minutes of Google searching found her the name and number of his station's contact point. It took several calls and some heavyweight name dropping to get what she needed – after thirty years in the force, Billy had three degrees of separation to officers even in neighbouring forces. Especially the old fogies.

It was almost half eight before Billy could place the call she needed to make.

And then, of course, it went to voicemail.

'*Sergeant Gavin Glynn. If you're calling about an emergency, please hang up and dial 999. Otherwise—*'

He would probably be involved in the terrorist incident. Billy had almost forgotten about it since the car bomb stuff subsided. Even if he'd been off duty, he would be back on duty and busy.

When the beep came, Billy took a deep breath and recorded a message explaining her need for an urgent response.

After she repeated her number, and left Dom's number too in case the mobile network went down again, she hung up. Her hands dropped into her lap. Another dead end.

Maybe it was the jet lag kicking in, but she felt shivery and exhausted. A kind of emotional hypothermia, the sort of tiredness that makes you stop fighting. She eyed the black coffee, but her bladder told her that even one more gulp could be the straw that made the camel wee itself.

Once it was dark, she might be able to sneak into the scraggly trees . . .

She glanced up and down the carriageway for a likely spot, but there was hardly any cover. Her gaze came to rest instead on the gaudy livery of the estate agent's Range Rover.

I'll take you home!

If only.

She noticed the livery. Then really noticed it. *Doors & Co.* Good name for an estate agent. But it was the phone number that caught her eye.

The dialling code was all wrong.

It had been right in front of her since she arrived on the motorway. Carl must have sensed her scrutiny because he jumped and looked all around, even checking his mirrors.

Billy got out of her vehicle, passed the clutch of drivers who shared significant looks in response to her sudden burst of action, and starting dialling the number on the estate agent's car. She meandered towards him as the call connected. The dialling code of his office phone number was the same as Dom's. But Dom's police station was fifty miles away from here. No estate agent works a patch that size. Carl Dawes had told Billy that he was here for a viewing, but he must have been lying. And not for the first time.

Chapter Thirty-Four

8.31 p.m.

Billy walked past the estate agent's car and held eye contact with the man inside while she waited for someone to answer the phone she was calling. But when the line connected, it went to an answerphone. Of course, the office was closed on a Friday evening. It listed an emergency out-of-hours number for tenants and landlords, so she tried that. By the time that number was answered, she was standing on the grass verge again.

'Doors & Co, can I help you?'

'I'm trying to reach one of your agents, Carl?'

'Mr Doors is unavailable now. Can I help?'

Something in her mind clicked.

Carl was Mr Doors. Not Mr Dawes.

She'd assumed Doors & Co was a pun, not a weird bit of nominative determinism. Presumably, it meant that Carl owned the company. But he'd given her the impression that he was just one of the agents. If this dialling code was anything to go by, his office was far from here, too far for him to be selling a house in this area, as he claimed.

'I'm Mrs Doors?' The woman nudged.

'It's Mr Doors I really need to reach.'

'He's on leave today so he might not have picked up his messages, but I can get him to call you tomorrow?'

'Maybe I can take his mobile number and call him in the morning?'

Mrs Doors gave the number and Billy tapped it into her phone.

'Before you go . . .' Billy said.

'Yes?'

'A friend is looking for an estate agent. What areas do you cover?'

Mrs Doors named a few places that were nowhere near this motorway.

Billy thanked her, rang off, and immediately dialled the number she'd been given for Carl Doors, the man who was fifty miles from where he should be. She strolled up the hard shoulder past Lincoln Quick – his lowered eyes still seeming to watch her pass – to reach her own vehicle. She opened her passenger door and then the glove compartment. The phone number Mrs Doors had given her should presumably be the burner phone that Carl had retrieved off the hard shoulder; he'd told Billy that his office insisted on giving its agents a work phone – so it should be the burner that rang.

Inside the poo bag, the burner remained silent.

But out of the corner of her eye, she saw Carl pick up his personal mobile. Billy marched back towards him as he contemplated the number and then pressed it to his ear.

'Hello. Doors & Co, Carl Doors speaking.' His voice was too animated, like a self-help guru. Billy saw his lips move behind the windshield but heard his voice down the line. A bit like speaking to someone who was being held on remand in prison, those booths with telephone handsets on either side of a Plexiglass window.

'Can you step out of your car, please?' Billy spoke pointedly into the receiver. His eyes raised to meet hers. Instead, he wound down his window. Billy was about to insist that he got out of the vehicle, but figured it might be better if he couldn't make a run for it.

But she was aware that with the large space in front of his car – which he claimed had been occupied by a motorbike – he could feasibly drive onto the hard shoulder and flee the scene. Not that he'd get any further than the blocked tunnel up ahead, but she didn't want him careering down the motorway.

She found a recording app on her phone and hit the red button. She gave her name, date and time, then informed Carl that she was recording.

'Carl Doors, are you the owner of the estate agency Doors & Co?'

'Along with my wife. Why?'

She cautioned him, his eyes glazing over at the *you do not have to say anything* part.

'Mr Doors, why did you tell me that your *boss*' – Billy bunny-eared her fingers – 'made you use a burner phone?'

'I just meant the management.'

'But you're the management?'

'Yes, yes, it's our policy. Safety, you know.'

'Very commendable.' Billy watched him. He held her eye, then swallowed and glanced down. Looked back up again. 'But it was a lie, wasn't it, Mr Doors? Why did you lie about the burner phone?'

He shook his head and said nothing.

'Carl Doors is shaking his head. Your office gave me your mobile number but it's not for the burner phone, it's for your own smartphone. So the story about the burner phone was a lie.'

'I use my own phone, that's true. But no one's going to attack me, are they? It's the women we worry about, going to viewings on their own. Suzy Lamplugh, you know.'

'Don't use the name of a murdered woman in your excuses, Carl.'

He said nothing.

'You had no property viewing in this area today. Your office just told me that you're on leave. This isn't even your patch. So why are you here?'

Nothing.

'I'm going to need your phone.' Billy held out a poo bag and after a long moment, he dropped the mobile in it. Now she had both his phones. 'I'm going to need your PIN number.'

He told her.

'And your car keys.'

He took them out of the ignition and dropped them into another bag.

'Don't go anywhere, Mr Doors.'

Billy went back to her car, already dialling Dom.

'Has there been a development?' Heathcote stepped into her path.

'Not now.' She performed a handoff that would have made a rugby player proud.

Billy sat in her own passenger seat and shut the door. Dom picked up and she updated him while navigating to the call history on the mobile he'd just given up. It was awkward through the thin plastic of the bag but after a moment, she said, 'He's wiped it.' Under 'Recents', Billy saw her own number – because she'd just called him a few minutes ago – but then nothing for weeks.

'He's deleted his calls.'

'Messages? WhatsApp?' said Dom.

Billy made a few taps on the screen. She was hardly a techie but knew the basics. 'Wiped and wiped. He's hiding something.'

'He's done it in a rush,' said Dom. 'Sitting in his car, panicking. "Clear All" is hardly subtle, it's the tech equivalent of taking a can of petrol to your filing cabinet. I checked the system, he's got no previous, but you need to be careful, Bill.'

She checked her rear-view mirror. 'He's sat in his car. Staring into space.'

'Leave him there until we have more. I got something on the motorbike.'

'The courier? I was just thinking that maybe Carl Doors made that up, but I definitely remember hearing a bike myself, so there must have been one . . .'

'There was. I rang round a bunch of courier firms, got a list of drivers with that model. Got a couple of plates. We picked him up on the ANPR at the next junction along. He must have cut through the traffic somehow. I have a call in to the registered owner, who didn't pick up but I'll keep trying. Local address. Seems legit, not even a speeding ticket, but the biker might shed some light . . .'

'Thanks, Dom,' Billy said, but she was already thinking about something else. The burner phone on the hard shoulder. Her mind rifling through scenarios like a hand rummaging through a messy drawer.

'I thought you'd be more impressed by my old-fashioned leg-work,' said Dom.

'Why are you trying to impress me?' she asked. 'I thought I'd lost my mojo.'

'Now I'm worried about what you might do to get it back.' A sigh rattled the line. 'I don't like you being there on your tod. It's getting dark, there's no backup—'

'I've got backup. There's an ex-con who knows how to use her fists, a posh lad who could do with some sense knocking into him, and his girlfriend is like a beautiful Jack Russell.'

'They are hardy little dogs.'

'Exactly. My fear is the traffic moving and they'll all be gone.'

'So what have we got on this estate agent?'

'It's only circumstantial . . .' But then Billy stopped. She had been toying with the phone while she spoke, navigating back to the calls section and all those deleted messages.

'The techies may be able to retrieve the data later,' said Dom.

'Doesn't help me now.' She tapped a different tab and a list of numbers, highlighted in red, appeared on the screen. 'Oh, whoops,' said Billy. 'He only deleted his received calls. Didn't delete his missed calls.'

She used her own phone to photograph the screen and pinged the picture to Dom. 'Here's some more legwork for you. Trace these numbers.'

'I'll put my best man on it,' he said. 'That means me.'

'I knew that already.'

Billy hung up then went through her notes and found the number of the burner phone . . .

There.

Carl's iPhone had missed an incoming call from the burner just after 4 p.m. That was weird – why did one of his phones call the other one? At that time, Lincoln Quick was still going through the airport. He had less than an hour to live.

Billy texted Dom and asked him to check the ANPR around the airport for a vehicle matching the estate agent's Range Rover. Should be easy enough to spot.

In all the excitement, Billy had failed to notice the sun sneaking under the covers of the horizon. The carriageway sinking into twilight. Some drivers had put on their interior lights. The illuminated cars gave off the bizarre festive glow of a Christmas market.

She winced and pressed the heel of her hand into her side, the muscles of her lower belly clenching around her bladder as though trying to comfort it. Maybe she could do a wee now it was getting dark?

Honestly, how much worse could people think of her after confessing to that whole scooter thing?

Chapter Thirty-Five

8.45 p.m.

'There are numerous levels of degradation,' said Pat philosophically. 'And a thin line between shame and intimacy.' She was holding a towel so that Billy could squat behind it on the grass verge. They were beside the black sedan. At least Lincoln Quick wouldn't peep. Which was good because the towel provided more of a mental barrier than a physical one. It covered little more than her embarrassment.

'Did he do it, then, the estate agent?' asked Pat. 'He was driving right behind the dead bloke.'

'He's helping the police with their enquiries,' said Billy.

'So he did it,' she said.

'That's not what that means.'

'Anyone with a prissy little beard like that deserves to go down. Give him a year inside and he'll look like Cast Away. Half the women did.'

Billy's belt had got caught up with her sleeve and she could only get her jeans to mid-thigh. 'Mother-of-truckers!' She wrenched the fabric down, spread her heels for balance and crouched as low as she could persuade her hip muscles to go. Then she waited. Her gut sent out shooting pains like lightning before thunder. She'd held it for so long . . .

And then she let go.

It was as close to a religious experience as Billy got. She dropped her head back to share the moment with the sky. A shooting star would have been appropriate. Or clouds parting to bathe her in moonlight. Instead, there was the golden halo of city lights and that was enough.

'There it is,' said Pat. 'Better now?'

'So much better,' said Billy, her voice still cramped. She tried to think of a more satisfying moment in her life, but came up short, which made her think – inappropriately and sadly – of Dominic's plea for her to 'go all in' and 'take a risk'. Was this life really the best she could do? Was it too late to hope for more? But then Pat moved and the towel was no longer covering Billy at all. The epiphany passed and now she was just a police officer peeing on a grass verge in front of a load of suspects.

She stared ahead willing the humiliation to be over. Stared at Lincoln Quick's black sedan. From the lower vantage point, in full squat, her thighs on fire and her stream sputtering, she could see the underside of his bumper . . . A vehicle on the other side had its headlamps on and the beam came under the central reservation to surround the sedan with light. Maybe here was the real epiphany. Something was hanging down behind the front bumper of the dead man's car.

Billy scrabbled for the wet wipe provided by the nurse, got herself sorted, pulled up her trousers and fixed her belt as she strode the few paces to reach the vehicle. Then she got down on her hands and knees beside the passenger-side wheel.

'Thank you!' muttered Pat.

'Thank you, Pat,' said Billy. 'Your assistance has been much appreciated.'

'What you doing now, then?' Pat squatted beside her, blocking the light from the headlamp.

'Move aside, there's something here.' Billy got her phone out of her pocket and switched on the torch. It illuminated the area underneath the front bumper. When Billy had scanned under the car earlier, she'd been checking for someone lying there, a quick glance. She should have looked harder. A piece of plastic was hanging down, hidden behind the bumper. She wouldn't have seen it when standing up, but from this angle it was obvious. Billy took photos of the plastic bit and then wiggled it free. It was a clip of some kind, part of a bracket. Now she inspected the bumper itself. There was a deep crack, a fault line running through it from top to bottom. A sign of an impact. The bumper had done its job of protecting the vehicle, but the impact must have snapped the bracket.

'What you got there?' Pat asked.

'This is a crime scene, Pat,' said Billy. 'I'm going to have to ask you to step away.'

Without a word, the woman did.

There was a six-foot gap between the black sedan and the car in front – the minivan belonging to the mother and daughter. A normal amount of space for two vehicles that had come to an abrupt stop in traffic. Close, but not touching.

Billy cursed herself. She'd assumed that the cars hadn't touched, that the black sedan had rolled to a halt – because its handbrake wasn't engaged, and the traffic had been moving very slowly, and because that was how she herself had come to a stop, just rolling . . . And because Kerry whatever-her-name-is hadn't told her any different.

Billy shone her light under the rear of the minivan itself. It was not a new vehicle, and not clean either. There were smudges and smears of grime. But on the left-hand side – right where the sedan would have tail-ended the minivan if, indeed, that is what happened – there was a corresponding dink. She switched off her torchlight. But the abrasion remained visible, lit by another light source.

Billy strained to look over her shoulder and saw a single headlight jogging up the hard shoulder.

A police motorbike?

At last. She was both relieved and disappointed – she'd come this far alone. Almost alone. She wanted to see it through to the end.

The bike was still a way off. She stayed down on the tarmac for a moment, looking up at the embers of the sky. The sedan hadn't just bumped the minivan, it had hit it with enough force to crack its own bumper. So why hadn't Kerry or her odd daughter mentioned it? They must have felt the impact. And Billy had asked them if there was anything untoward – getting tail-ended by a dead man would certainly qualify.

Time to talk to Kerry and Hyacinth again.

Billy shifted onto her knees and was about to stand up when a rush of noise and air told her the motorbike was coming faster than anticipated. There was no way a police officer would drive through traffic at that speed. Billy hit the ground underneath the minivan as the bike roared past. Missed her by a hair.

She rolled onto her feet and shouted after it. There were people all over the carriageway, the idiot would kill someone at that speed in the dark.

Its brake light glowed red. The engine revved. And then a white light shone in her direction. The driver was coming back. Moments later, it drew up in front of her and stopped at an angle on the hard shoulder. She recognised the description of the driver. No leathers. Courier bike. Box on the seat.

It was the Husqvarna Nuda 900.

Chapter Thirty-Six

8.50 p.m.

'You Sergeant Billy?' asked the muffled voice of the biker.

She agreed that she was.

He removed his helmet and a satin beanie that covered twists. He took off his gloves too, and held out one hand to shake. Then he half withdrew it, as though concerned that shaking hands with a police officer wasn't the right thing to do.

'Your man called me. Said you had a few questions. I thought it would be best to come back here. I didn't want you to think I'd left a crime scene, thought that didn't look too good for me, I don't want no trouble with the police.'

'No trouble, I appreciate you coming back. You did well to find me.'

'Your man gave me your three words so I found your location.'

Billy felt itchy in her fingertips. This investigation was like a parcel that had been overwrapped. It was tempting to tear it apart. But instead she ran details over in her mind, picking at them like fingernails feeling for an edge of Sellotape. The end was here somewhere. Maybe this biker could help unravel it. Billy decided to speak to him before confronting Kerry Wells and her daughter, Hyacinth. The whole time they'd been on the motorway, those two had been the least sociable. It hadn't seemed odd; they had each other, while everyone else needed company. And the girl seemed introverted to say the least. Billy didn't want to slap labels on her, but Hyacinth must be around the same age as Daisy and the two young women were chalk and cheese. Perhaps it was this unconscious comparison that made Billy aware of Hyacinth's inadequacy; while Daisy was bright, Hyacinth was colourless.

Billy sent a text to Dom; photos of the two damaged cars and her theory that the sedan had tail-ended the two women.

Then she turned her full attention to the biker.

He introduced himself as Leon Musa, thirty-one, a courier. He wasn't wearing leathers because his trousers were made of a special fabric that looked like denim but contained Kevlar.

'They're as expensive as leathers,' said Leon, brushing a fleck off a pleasant thigh. 'But you can wash them. Some of the other bikers stink.'

'Is this where you stopped?' Billy indicated the gap between the estate agent's Range Rover and the black sedan.

'Hard to say, to be fair. I thought I'd remember it more clearly.' Leon seemed less confident now, as though he'd volunteered for something without thinking it through. He looked around. Gave the estate agent a long, hard stare. Carl Doors was gazing at the grass verge like it was a TV screen. Away with the fairies. Leon looked at the back of the black sedan.

'That the dead guy?'

Billy nodded.

'Thing is, I stop and start all the time on the bike. I have to be more vigilant than car drivers.'

'If I had a pound for every biker I've scraped off the tarmac . . .' she said.

'If I had a pound for every time some idiot's pulled out in front of me. Or thrown open a door as I come past— Oh!' Leon stopped. He turned on the spot, noted the camper van and ticked his finger. Billy couldn't work out if he was genuinely trawling his memory or if this was a bit of acting for her sake. Then Daisy got out of the back of the camper van.

'I remember her,' said Leon. 'She gave me a Paddington Bear stare. In fairness, I nearly took off her wing mirror and I was a bit late so I was giving it some throttle. Must have been loud. I glanced up as I went past, just at the moment I thought I was going to hit her mirror, you know? But then I just tipped it an inch.' Leon mimed with his shoulders using his weight to shift the bike a fraction to the right. 'I remember thinking, that girl's peng but what a nasty frown.'

Billy forced herself to say nothing. Daisy had said that the courier bike came past her window, even though that was an awkward route for it to take. Billy wanted confirmation. And an explanation for why he went that way. But she couldn't feed it to him.

'Yeah,' decided Leon. 'This is where I stopped. This is it. Don't remember that sedan.' He waved a finger at Lincoln Quick. 'Or the SUV.' He indicated Charlotte's car. Then held his right hand up like a cocked gun. 'But I remember her' – Leon pointed a finger forward at Daisy – 'and him in the Rangey,' a thumb backwards at the estate agent. 'He was driving like a cock and she was the pretty one.'

'What was wrong with his driving?'

'Overtaking then lifting off the speed. I go past him then he races past me, cuts in, slows down, rinse and repeat. Irritating. We're going round and round each other like a pulsar, you know? And then we stop and he's right up my backside again . . .' Leon glanced over his shoulder. 'He looked about as with it as he does now.'

'Did you see him get out?'

'Of the car?'

'Yeah, after the traffic stopped.'

He squashed his mouth down.

'Nup. Don't think I looked back after the traffic stopped.'

'So what did you do?'

'I don't have to sit here in a gridlock like a lemon, do I? So I left these lot making lemonade and I went on my way. Work to do. That's not illegal, is it?'

'Depends how you go on your way.'

'I didn't drive down the hard shoulder. I made that mistake a long time ago, got a ticket.'

'Are you sure?' Billy held up both palms. 'I'm not going to ticket you now. I just need to know your exact route.'

Leon laughed, more relaxed now he was orientated. 'God's honest truth, every now and then I may go on the hard shoulder to get round one car if there's an obstruction. But on this occasion, I went past the pink Micra. Like I said, I nearly took off her wing mirror, it was tight getting between her car and the central reservation.'

'So why did you go that way? Doesn't seem like the obvious route.'

'As you can see, this Audi is so damn wide there's no space either side. I hate SUVs. And the hard shoulder was blocked, so.'

'Blocked how?'

'If I remember correctly,' said Leon, 'one of the cars had its door open. Must have been that black sedan, the one with the dead driver.'

Chapter Thirty-Seven

8.55 p.m.

Twilight had dimmed the long view up the motorway, the vehicles outside the grid reduced to a sprinkle of lights like a city scape. It was dark enough that if someone wanted to do a runner, now would be the time. Not that it would do them much good, Billy had a record of who was who. The killer must know too that the moment for escape had passed. Which raised a question; why stay? When the traffic came to a halt almost four hours ago, they might have figured that running off would give them away. But now the killer had nothing to lose. So maybe they had a plan. Maybe they were too scared. Either way, as Billy's circles grew tighter, the killer got more hemmed in, more dangerous.

Leon Musa agreed to stay until more officers came to take statements. Billy walked towards the Toyota minivan. Kerry Wells sat in the driver's seat alongside her daughter. Hyacinth's hand curved over her mouth as she gnawed the thumb.

The wind fizzed through the trees. It came down the carriageway like a wind tunnel, channelled by the high walls on both sides. Billy retied her hair higher on her head, to keep it out of her face.

She dialled Dom's number. While she waited for him to pick up, she walked to the back of the van whose rear window was decorated with decals from tourist sites. An ammonite from a fossil museum. A thistle from a Scottish castle. A spider's web from an insect centre. Billy placed her fingertip on that one. If this case were a web, this vehicle was the point where all the threads crossed. She kept her finger in the web, but her eye moved up to a car dealership label; Right Price Autos, Toyota specialist, Howle Green.

Bloody Howle Green.

That village again.

Dom wasn't answering so Billy hit the red button.

Kerry didn't live in Howle Green, did she? Billy checked her notes for the address where Hyacinth still lived with her mum. She typed the two places into Google Maps. Howle Green and the suburb where the women lived were twelve miles apart. There must be dealerships closer than that.

Billy's phone rang. Withheld number. She answered at once.

'This is Sergeant Gavin Glynn, I picked up your message.'

She ran through the situation. She needed to know exactly what happened on the night Lincoln Quick killed Thomas Kingston. The real story, not the stuff in the press.

'Not a lot to tell you.' His voice was high and strained, shouting over a roadside racket. He must be at the scene of one of the incidents in the city. 'I was first on the scene after the patrol car, but once it turned into a diplomatic incident, it got passed up the chain of command faster than the bar bill from an office party.'

'Between me and you,' said Billy. 'I'm looking at Lincoln Quick now. He's dead. With a bicycle spoke in the back of his neck.'

'A what in his neck?' The yowl of a siren in the background obliterated his voice.

Billy gave it a second to pass.

'A spoke from a bicycle wheel,' she said very clearly.

'Interesting,' said Gavin. 'But if it's got anything to do with the Americans at RAF Howle – or the British Secret Service, for that matter – you might as well let it go. Hand it over to the first suit who arrives on the scene and go home. They have zero respect for the likes of us plod.'

Billy could imagine his lips spitting on the receiver.

'I need to know about that night. Whatever you can remember.'

'Ah, well, it looked like a simple hit-and-run, albeit a little unusual as it involved a bicycle. But as soon as we realised one of the RAF Howle lot was involved, one of the spooks, it flew up the chain of command. Thomas Kingston went off in an ambulance. I was sent away from the scene. After that, I was learning about it from the newspapers, like everyone else.'

'This murder on the motorway connects somehow to the accident in Howle Green.'

'Accident?'

'Wasn't it?'

'Hard to say without ever getting the chance to question the cyclist. Did you see the CCTV?'

'There's CCTV from that night? Did it look like an accident? Or more than that?'

'All I can tell you is what's on the cameras. Lincoln Quick on his expensive bike. He goes through a red light *and* he was riding on the wrong side of the road. Thomas Kingston reaches the crossing and looks to the right – the correct way, the direction that any traffic should be coming – so he doesn't see a bike coming at him the wrong way. It was over in seconds. Kingston gets hit, rider comes off his bike. But then Lincoln Quick has a good hard think about it, and decides that, yes, he is a total shitbag and will happily leave a man to die in the gutter, and he rides off. All that's on camera.'

Billy thought about that for a moment. 'Did Lincoln Quick know the victim?'

'Thomas Kingston?' said Sergeant Glynn. 'Not as far as we know.'

'So what makes you think it was more than an accident?' Billy asked.

'Accident, negligence, recklessness . . . Was the Yank on the wrong side of the road because he had a brain fart, or was he drunk, or did he intend to ram the guy? Who knows. Because we never got a chance to question him. He should have called an ambulance, that's for sure. Instead, he fled the scene, fled the country. Thomas Kingston died alone in the street. And there was nothing we could do about it.' Glynn's voice shook a little.

'What happened to the CCTV?'

'I've got a clip, if you want to see for yourself?'

'Yes, please. What about Thomas Kingston. Who was he?'

'Normal bloke. No reason to get murdered. Small accountancy firm, did alright for himself, not flash. Wife and daughter. They liked to travel. Nothing to suggest he had any beef with Lincoln Quick. Nothing to suggest he'd been targeted specifically. Or had anything to do with spooks at RAF Howle. So maybe it was just an accident. But it was an accident that wouldn't have happened if Quick had been on the left-hand side of the road and hadn't gone through a red light and hadn't been riding too fast. And Kingston might not have died if Quick had stayed at the scene and called an ambulance instead of saving his own arse by heading back to the base like a kid running under his mother's skirts.' Sergeant Glynn ended his speech sounding heated enough to show there had been considerable emotion expended over this case.

'And Quick was bundled out of the country that night under diplomatic immunity?'

'Irony is, there isn't even a law in England that covers dangerous driving on a bike.'

'I remember reading about that. There's only some old-fashioned law?'

'Death by dangerous driving covers motorised vehicles. For a bicycle, you would have to use a law that dates back to horses and carriages. Furious driving, it's called. So he probably would have got away with it. But he legged it, the Americans gave him diplomatic immunity, and now there's been a warrant out for his arrest. Or there was.'

'Any idea why he would come back to England when there's a price on his head?'

'Because he's a silly fucker?'

'You mentioned that Thomas Kingston had a wife and daughter?' Billy took in the tourist decals on the back of the van. *They liked to travel.* 'I'm asking because the vehicle I'm interested in right now on the motorway contains a mother and a daughter, and a sticker from a car dealership in Howle Green on the back window. But the mother isn't called Kingston . . .'

'I did the death call. It was Kerry something.'

'Kerry Wells.' Billy's heart gave a lurch. 'And the daughter?'

'Grown up, but seemed a bit . . .' Glynn made a humming noise. 'Not all there. She'd never left home and was much older than she looked.'

JO FURNISS

Billy checked the details in her notes. Dates of birth. Yeah, Hyacinth was in her early thirties. She looked the same age as Daisy but was ten years older.

Gavin was still talking: 'Kept mouth-kissing her dogs. It was, yeah, I put it down to the shock, but . . .' He clucked his tongue and let the implication hang.

'Was she called Hyacinth?'

He groaned. 'That's right. Awful name – that programme on the telly, do you remember?'

Up ahead, the gantry lights went out.

'Hold on, something's happening,' Billy said.

'You know they've lifted the cordon around the city centre?'

'Shit. Then this traffic could move.'

The gantry illuminated again with red crosses.

Billy kept her voice low. 'I think it's them, Gavin. Kerry and Hyacinth. In a minivan on the motorway, parked right in front of the man who killed their husband and father.'

'Consider them dangerous,' said Glynn. 'I'd back off until I can get a unit out to you. The higher-ups are going to be all over this.'

Billy's phone pinged. An image with an arrow on it. A video clip.

'Just sent you the CCTV,' Glynn said. 'I'm calling it in to my super. She'll run it up the chain of command.'

'I've got enough to talk to the two women right now.'

A pause.

'This isn't your patch, Sergeant.'

'The traffic will move and we'll lose the crime scene. I'm going to talk to them.'

'Sergeant—'

'Get here as soon as you can.' Billy hung up and turned her phone to airplane mode. She pressed play on the CCTV clip. The video had no sound. It only lasted a few seconds. But it made her colder than the night air.

The camera is set up perpendicular to the road so that it looks over the crossing. Low light, early evening, a man – Thomas Kingston, presumably – is eating a bag of crisps as he approaches the junction. One hand reaches to press the button and the walking man symbol illuminates right away. He takes two strides before a blur appears from his left. Then the crossing is empty. Only the crisp packet on the ground.

It took Billy a moment to realise that the video was still running.

Movement at the very edge of the shot. A shadow cast by a person getting up off the road. Then he staggers back into shot. A slight figure in cycling gear. No helmet. Maybe it came off in the crash. Maybe he was too cocky to wear one.

The image was grainy but it was recognisably the man in the black sedan. Lincoln Quick.

He stares towards the side of the road where Thomas Kingston must be lying out of shot. Dying. Then Quick clutches his head as though his skull might explode. He turns to look down the road, back the way he came. He stands there for a long moment, hands over his ears. His face contorts, a grimace or swear words. Then his posture changes. He drops his hands and pulls himself upright. He looks up and down the road as though checking for traffic or, more likely, witnesses. Then he walks away. A second later, a bike swings round, flashing briefly in and out of the shot. And the road is empty again.

The clip stopped.

Billy scrolled the video back and watched the last few seconds again.

Lincoln Quick turns to look down the road, back the way he came. He stands there for a long moment, hands over his ears. His face contorts. Posture changes. He stands upright.

What happens in that moment, the moment he decides to run?

It had been a life-changing decision for Quick. A life-ending one for Kingston.

Inside the minivan, Hyacinth Wells stared straight ahead, sucking her thumb. Oblivious to the fact that Billy was watching a video of her father's untimely death. Billy felt a pang for the girl. No wonder she appeared strange, traumatised. If Billy's father had been mown down at a crossing and left to die in the road, she might feel murderous too.

From some car radio in the vicinity, drums signalled the news. Billy took in enough of the headlines to learn that the siege was indeed over. As soon as there was a lane free through the Deadwall Tunnel, her crime scene would literally drive off into the sunset.

THE FIFTH HOUR

Chapter Thirty-Eight

9.01 p.m.

Darkness had crept up over the edge of the carriageway to sur-round the grid of vehicles around the black sedan, giving the scene the air of a stage play. Interior lights glared. Billy wanted to clap wildly to break the spell, except it wasn't an act. It was a crime scene.

She stopped beside the passenger window of the minivan. Hyacinth's hands dropped into her lap. Her thumb was wrapped in a plaster, torn into ragged peaks by her teeth. It was a child's plaster. Decorated with a picture of Batman.

A vigilante.

Billy tapped on the window, making Hyacinth jump. The young woman flapped a hand under her chin as though fanning herself. Billy pointed at the side door of the Toyota Sienna.

'Why?' asked Kerry.

'I'd like to see inside,' Billy said.

A moment's hesitation. Then the door clicked and zipped along its rail, power assisted, operated by a button on the dash-board. That would have made it easier for someone to slip inside.

Billy peered in but stayed on the hard shoulder. If Hyacinth had moved from the black sedan into the Toyota, then vital fibres from Lincoln Quick could match him to this vehicle. She didn't want to disturb that evidence or be accused of transfer-ring fibres herself.

'You can shut the door,' Billy said. It rasped to a close.

She stood instead by the passenger window, which Hyacinth slid down until it was fully open. Kerry crunched around in her seat to face Billy. Hyacinth only gave her a side-eye.

Kerry's face took on a constipated look as Billy started recording on her phone and delivered the warning that anything she said could be used against her in court.

'Mrs Wells, did you notice anything unusual as the cars came to a stop earlier?'

'You asked me that before. Anything untoward, you said. And I said no, there wasn't anything.'

'Nothing strange?'

'Like what?'

'Anything that might have got your attention?'

Kerry threw her hands up and widened her eyes.

'Mrs Wells is indicating that she doesn't know,' Billy said. 'Did the vehicle behind yours, the black sedan, make contact with your car?'

Pause. 'I don't think so.'

'You don't think so?'

'It's hard to say.'

'Is it?'

'Yeah, in traffic, when it's stop-start, it's jerky.'

'But you'd know if a car bumped into you?'

'Maybe.'

'Definitely,' said Billy. 'You'd feel a jolt.'

Kerry hummed as though they were discussing a preference in the flavour of jam and Billy had said quince. Like it was an unexpected opinion, rather than a statement of the bleeding obvious.

'Have you ever been tail-ended by another vehicle?' Billy asked.

'No,' said Kerry. 'Well, yeah, but years ago.'

'Me too,' said Billy. 'It was weird, because we weren't travelling all that fast, but I felt this jolt and then I realised the rear-view mirror was skew-whiff. The driver in the car behind tried to claim he hadn't made contact, but the jolt was obvious to me. And there was damage on the bumpers of both cars so there was no disputing it in the end. His insurance paid up.'

Kerry's face was so still, it was like a movie that had been paused.

'I think you noticed that the vehicle bumped you, Mrs Wells.'

Kerry said nothing. Hyacinth was watching her mother.

'And I'm wondering why you didn't tell me that?'

'I didn't.'

'Didn't what?'

'Didn't notice.'

'So are you saying that it didn't hit you or you didn't notice that it hit you?'

'Both, I suppose.'

'They're two different things.'

'It didn't hit me.'

'Even though there is visible damage to both the vehicles?'

Kerry gave a shrug so small it might have been a twitch.

'What about you, Hyacinth?'

'Didn't notice,' she said, surly as a teen even though she was over thirty.

'How could you not notice the car behind running into you?'

'Maybe it didn't,' said Hyacinth.

'But it did,' said Billy. She held up the photo of the broken bracket on the black sedan. 'The bumper of the car behind is damaged.' She flicked to the next image. 'And so is the bumper of your car. We'll match the paint when we take the cars in for investigation.'

'I'm not lying,' said Hyacinth.

'So did you notice the prang or not?'

Silence.

'Hyacinth, did it hit you or not?'

'I didn't feel it hit this car.' The younger woman looked triumphant. Like a child who's come up with an excuse not to eat their vegetables.

'Maybe you didn't feel it because you weren't inside this car?' said Billy.

'What?' said Kerry.

Billy kept her attention on Hyacinth. 'You said you didn't feel it hit this car. You've chosen your words very carefully, Hyacinth. If you weren't in this car, then you wouldn't be lying, would you? You couldn't have felt it hit you if you were in the car behind, in the black sedan.'

Silence.

Kerry slid round to face the front.

'You never took your husband's name, did you, Kerry Wells? That's why I didn't connect you to Thomas Kingston. And from there to his killer, Lincoln Quick, who's dead in the vehicle behind us.'

Silence.

'What surname do you go by, Hyacinth?'

'She took my name,' interrupted Kerry. 'My husband said I should have the honour as I was the one who pushed her out. It got a few people's backs up in the family, but that's what he wanted. And he got to choose the name Hyacinth, after his mother.'

'Yeah, thanks for that,' the young woman said, tugging her face into a frown of mock indignation. 'Can you imagine if they'd called me Hyacinth Kingston-Wells? Sounds like a town. A spa town.' Her performance had the stale air of a repeated joke. She probably said the same thing every time. Billy knew fear when she saw it. Gallows humour, inappropriate reactions. Had seen it many times on the streets and in the nick.

'Kerry, you moved into Hyacinth's flat after your husband's death, is that right?'

'I wanted to get away. And we're selling the family house. Why?'

'Your car isn't registered to Howle Green anymore, but I see from the dealership sticker that you bought it there, so I'm assuming you bought it when you lived there?'

'Yeah.' Her tone implied a silent *and?*

'Who can blame you for making a fresh start after Thomas was killed by an American who skipped the country.'

'Tommy,' she said. 'We called him Tommy.'

'Would you remove the plaster from your thumb, Hyacinth?'

'What for?' said Kerry.

'You've been sucking your thumb and chewing on it ever since the traffic stopped. Now you've got a plaster that you didn't have before. I think you've got an injury. A puncture wound. From pressing the end of a metal rod with enough force to penetrate someone's neck and enter their brain.'

Hyacinth cradled the injured thumb. But she didn't remove the plaster.

'Claiming diplomatic immunity in order to skip the country is one thing,' said Billy, 'but returning to Howle Green is rubbing your noses in it. No wonder you're angry, Hyacinth. No one would blame you.'

'Blame her for what?' spat Kerry.

The air got thinner.

'How did you know he was coming to England?' Billy asked.

Silence.

'Maybe you got a tip-off. And it was a chance to get justice. Somehow, you got into his hire car at the airport and hid in the footwell, then stabbed Lincoln Quick in the neck on this stretch of road, which is always snarled up on a Friday and not covered by surveillance cameras at the moment – I wonder who gave you that information about the CCTV?' Billy shook the irrelevance away. 'Doesn't matter, we'll find out. I'm sure a lot of people are sympathetic enough to help you. I'm sure many people would have cheered you on. So you stabbed him, Hyacinth, as the traffic slowed down almost to a stop. The plan was to steer his car onto the hard shoulder so it looked like a breakdown—'

'Stop it!' wailed Kerry.

'—and for Kerry to pull over like she was helping. That way, no one else would bother to stop. Not on a Friday night. The other cars would sail on by, in a hurry to get home. Then Hyacinth would sneak out of the sedan and into the minivan and you could both drive off unseen. Except the terrorist attack caused gridlock. The traffic came to an unexpected halt and, instead of rolling onto the hard shoulder, the sedan bumped into the back of you and stopped in the inside lane. You were trapped at the scene of the crime. Hyacinth slid out of the sedan anyway – that's why its back door was open momentarily and prevented the motorbike from driving past on the hard shoulder – and then you got in here through the sliding door. And now here we are.'

Both women stared fixedly ahead as though their journey was continuing uninterrupted. Instead, the traffic was as solid as a brick wall.

In that moment, Billy felt quite sorry for them.

A rap on the roof made Billy jump and Hyacinth yelp.

Carl stood on the road beside Kerry's door. He tapped her window but she refused to open. He crouched down awkwardly to look right through the vehicle at Billy. His eyes were very wide above the anchor of his silly beard. She willed him away; she was in the middle of something.

'I've got something to say.' His voice was muted by glass. 'A confession.'

Chapter Thirty-Nine

9.07 p.m.

'Mr Doors, can this wait?' said Billy.

'I need to talk to you,' he said. 'I need to tell you—'

'Go back to your car,' interrupted Kerry.

His eyes slid away for a moment but recoiled from Kerry's glare and landed back on Billy.

They know each other.

If he knew Kerry and Hyacinth, and no one had bothered to mention that fact either, then she wanted to hear what he had to say.

'It's about the phone,' he said.

Billy walked around the bonnet to reach Carl. She waved him towards his own vehicle, and fished in her pocket for a new poo bag, which she opened and held out to Kerry.

'Car key,' she said.

Kerry took the key from the ignition and dropped it in the bag. Billy tied and pocketed the bag as she followed Carl.

'Can I remind you you're still under caution?' She set her phone to record.

'Can I remind you that I'm giving this statement voluntarily?'

'Duly noted.' For all the difference that would make.

'I didn't want to do it,' he said.

'Do what?'

'I have to start from the beginning.'

'Mr Doors, I don't have much time here—'

'If I'm going to tell you, I want to start at the beginning otherwise you won't understand.'

Billy nodded for him to carry on.

'Last year, I had a bit of a wobble.' He stroked his beard. Maybe the topiary was a symptom of said wobble?

'What kind of wobble?'

'A wobble in my marriage.'

'Your wife is a co-owner of the estate agency?'

'Yeah, we started the business fifteen years ago. Married the same year. Three boys, all in private school.' He said this as though it should earn him brownie points. 'You might say we've gone all in.'

Not another one.

Dom would approve.

'Right, and this wobble?'

'Dating app. You heard of Dizcreet?'

Billy gave the barest shake of the head.

'It's for married people. To meet anonymously. No strings, no consequences.'

'An adultery app?'

'I suppose so, yeah. I didn't want to do any harm, didn't want to end my marriage, just needed a . . . bit of spark. I was all work. Joyless routine. Miserable. So I went on Dizcreet. Only a couple of months. Met a few women for the odd night. You know . . .' His eyes were on the middle distance.

'Not really, but go on.'

'I ended up getting blackmailed.'

As you do.

'I got a phone call out of the blue a month ago.'

'One of the women?'

'It was a bloke.'

'What did he want? Money?'

'No, this is where it's weird. Well, it was all weird, you never think something like that could happen to you, but all of a sudden I was in the middle of this clandestine thing. It was like a Liam Neeson movie. He wanted me to run an errand – the bloke who called, not Liam Neeson. That's what he called it, an errand. At the airport.'

'Today?'

'Yes, a week ago he sent me that burner phone. I had to pick it up from one of those delivery lockers outside the supermarket. Then today I had to go to the long-stay car park and wait for a signal. I still didn't know what I had to do, but he said it wasn't illegal or dangerous, I just had to talk to someone. I mean, I'm not stupid, I knew it had to be ... something. Anyway, I sat there for ages, well over an hour. I thought maybe the burner phone wasn't working so I used it to call my own number and then called it back, which I immediately realised was stupid because now there would be a record between the phones.'

Indeed there would.

'But about two hours after I got there, he finally texted me a bay number and said I had to get talking to the bloke who arrived to pick up the car that was parked in the bay. It was a hire car.'

'Was it this hire car?' asked Billy, pointing to the black sedan.

'Yes, it was that car.'

'In which there is now a dead man?'

Carl swallowed. 'I didn't expect this.'

'You must have suspected something criminal? Why didn't you call the police?'

'Because my wife ... I thought I could get it over with. I thought he'd turn up and ask me for money, we'd drive around a few cash machines, or maybe he'd make me deliver drugs or something, like county lines and that. I don't know what I expected, I just didn't want my wife to know. I'd lose her, the boys, the business ...'

'And what happened after he texted you?' Billy asked.

'A bloke arrived about ten minutes later. Obviously just come off a flight. He had a small cabin bag. I'd been told to wait until he unlocked the car and put his case in the boot or whatever, although he actually put it in the front seat, and then I had to engage him in conversation for a few minutes.'

'How did you do that?'

'I pretended I'd dropped my car key and asked if he'd seen it, and he walked up and down the row with me, even got down on all fours to look underneath cars. After a bit I pretended to find the key behind a wheel, and then I asked him where he'd come from, you know – because by this time I was kind of curious to know what all this was about, why I'd been asked to meet this bloke who seemed perfectly normal apart from being American – and the bloke said he'd come from Washington and I asked if he was on business, and he said he used to live here but moved back home – and we had a quick chat. It's easy for me, the banter, I do it all day every day.'

'Did he say why he'd come back to England?'

'Just visiting, he said.' Carl shrugged. He looked relieved, as though the worst was over, and Billy didn't like to tell him that it almost certainly wasn't over because after all this came out his wife would definitely know. 'I can't believe he's dead. Bloke seemed normal. Bit stiff, shoulders back, you know, I thought perhaps he was ex-military.'

'He's a spy. CIA.'

Carl's face visibly blanched, his hazel skin turning a little yellow. 'Shit. The bloke on the phone said he wasn't dangerous.'

'He lied.' Billy nodded at the black sedan. 'But I'd say it turned out to be more dangerous for him than you, don't you think?'

'That's why I'm telling you. This was nothing to do with me.' Carl jabbed a finger at the sedan. 'I didn't touch him. Just talked to him. Then he got in the car and drove off, and I got in my car too.'

'What you doing here, then? You followed him?'

'Yeah, I . . .' Carl cast his gaze about the clouds, looking for inspiration. 'After I did it, I got this feeling in my stomach, like something was going to burst out. Anger, I suppose. I just thought, no! I am not having this. Not being held to ransom, some stranger threatening my wife and sons.'

'The person who called you threatened to kill them?'

'Threatened to *tell* them. About Dizcreet. He knew details. He was going to email my wife.'

'I think you had to distract the driver in that car park so that someone could get inside his hire car. Did you see anyone else there?'

'It was a big empty car park. Well, full of cars, obviously, but no people. A shuttle bus drives all the way around the outside every twenty minutes. A handful of people would get off at each bus stop, go find their vehicle, drive away. But there's acres and acres of car park. I was told to go to a specific area where the rentals are parked.'

'Was there a car hire office there?'

'No, nothing. Just a sort of telephone box, with an emergency phone inside in case there's a problem. My guy didn't have a problem. He found the car, I waited until he unlocked it with the beeper, then I went up and talked to him. And that was that.'

'And then you followed him?'

'After he got in his car, I went back to mine in the short-stay car park, paid my ticket and drove out through the barrier and I spotted him right away. He'd stopped in a lay-by after the roundabout, making a phone call or setting his GPS or something. And, like I said, I was so angry. Who was doing this to me, you know? He was my only clue and he hadn't seen my car so I thought I'd just see where he was going. We drove straight here. Nothing strange. Wasn't speeding or anything. Just driving. Traffic was awful on the motorway, but that made it easy to follow him. Except when that motorbike kept weaving about in front of me.'

'And when the traffic stopped?'

'It stopped suddenly and we sat here for a bit. The pretty girl got out first and was being all like' – he flapped his fingertips at his throat – 'drama queen. And then that woman in the Audi spoke to you. I don't know what you said to upset her, but she stomped behind the car, face like a busted clock, and she spotted the dead bloke right away. I already thought he looked odd, but obviously I couldn't get out to have a look in case he recognised me. But I didn't realise he was dead until she started screaming. Then I was shitting myself, sat here knowing I was up to my neck in it.'

'What about the burner phone?'

'I chucked it out of the window. Panicked. The guy I'd talked to was dead. Like, what the fuck? It was obviously something to do with the airport, and the blackmailer, and he'd sent me the phone using my name, so . . . It was like the thing was hot, physically hot, and I wanted it out of the car, just wanted it gone. But the moment I threw it, I realised that I was fucking stupid – I wasn't lying when I told you that before – I got out and picked it up. Only that bloody *unit* was watching . . .' Carl jabbed a thumb towards Pat who was standing on the hard shoulder in the exact same spot Billy had first seen her. 'I saw her, standing on the hard shoulder just like that with her arms crossed, foul look on her face.'

'Why did you wait to tell me all this?'

'Why do you think? Doesn't look good, does it? I only came here today to save my marriage, and now look, my wife is going to find out, isn't she?'

'She is,' agreed Billy. 'But this is a murder investigation. It's important.'

'So what? You're going to pin a murder on me? I didn't do that. It had nothing to do with me.'

'Unless you assisted a murderer in getting into the hire car.'

Carl's moist lips parted, his tongue nipping out for air.

'It is in your own interests for me to solve this case, Mr Doors,' said Billy. 'Because accessory to murder might get you three years in prison, but murder would mean life. You won't be seeing much of your wife and sons then.'

'Don't you think I haven't been sat in my car, going over this? I didn't see anyone at the airport.'

'Would you say you distracted the victim from the hire car? It was unlocked and unattended for a time?'

Carl thought for a beat. 'Someone could have got inside, yeah.'

'And what about here on the motorway? You had the best view. You told me you were leaning down into the footwell.'

'That was true. I was rooting out the burner phone, like I said—'

'Sergeant!'

Daisy's voice rang out at the same time that the engine of the minivan coughed to life. Kerry had somehow started the engine. She must have a second key fob.

The minivan lurched into gear and shot backwards, hitting the sedan and sending it rolling backwards into the gap. That was the end of her fender evidence. Billy and Carl jumped aside as the sedan stopped just shy of his Range Rover. Then the Toyota Sienna swung violently onto the hard shoulder and drove away.

Kerry and Hyacinth Wells were making a run for it.

Chapter Forty

Billy started running to her vehicle. She bellowed at Pat to shift her car out of the bloody way, then shouted instructions to Carl over her shoulder to make a list of all the women he'd met and the places he'd stayed. Her mind skittered faster than her feet. It could have been a data breach of the Dizcreet app that led to him being blackmailed. But it was more likely one of the women he'd met in person. Billy reached her car and flung open the driver's door, then hesitated; follow Kerry and Hyacinth, or stay with Carl? His blackmailer was involved somehow. But two people fleeing a crime scene demanded her attention.

'What's going on?' asked Heathcote, striding up to Billy's window as she landed in her driver's seat. She slammed her door in his face and started the engine. On second thoughts, if he was coming over to mansplain things again, he could make himself useful. She slid down the window.

'Remember this number,' Billy reeled off her phone number as she performed a many-point turn to get out from between the Audi and the Renault. The landlord dithered, repeating the number but getting lost after a few digits. Daisy appeared with her mobile in hand, 'Say it again?'

Billy repeated her number, then said, 'The estate agent is going to give you a list of names and places, I want you to text it to me. He doesn't have a phone. I've taken it.'

'On it,' said Daisy.

Pat had shunted forward onto the hard shoulder and her tail-lights lit up white as she reversed to get out of the way and leave the lane free for Billy. She hit the accelerator, feeling her tyres spin on the hot tarmac.

Take it easy, Belinda.

She eased off and the unfamiliar car gained traction. She manoeuvred through the tight space, bumping onto the grass verge and back onto tarmac to follow the minivan.

Its taillights were way ahead. Where were they even going with the Deadwall Tunnel still closed? She squeezed all she could out of the hire car, but it was a shitty rental, bottom of the range, the pedal flabby under her foot. The minivan was hardly a high-performance vehicle either. And it couldn't go anywhere except into a bottleneck. Billy only had to catch up. There was no need for a police chase. That lesson had been learnt. No need for another accident like the scooter. She eased off the speed.

But then the lights of the Toyota vanished. Disappeared. The lane ahead was dark, lit only by pools of illumination from stationary cars in the inside lane.

Kerry had turned off her lights. Gone dark.

So dangerous. So desperate.

Billy kept her eyes fixed on the lane ahead, terrified that a bystander might step out and she'd run them down – a child, maybe, that stray dog – another freak accident that Billy would be responsible for.

If only she had the blues and twos of the patrol car.

She flicked her lights onto full beam and jammed a fist against the wheel, sounding the horn continuously as she drove into the darkness. They would hear her coming. But please, God, don't let anyone step in front of the dark Toyota minivan.

Up ahead, the shape of the vehicle flashed through the ambient light that pooled around the stationary cars, a dark hulk moving in the shadows. Billy thought of Batman again, the vigilante on Hyacinth's injured thumb.

They drove for over a mile, vehicles sliding by Billy's window on the right-hand side, a kaleidoscope of shocked faces.

And then a flash of light on the left. Beyond the grass verge, where there should be only darkness.

She hit the brakes and fishtailed to a stop.

There had been a gap in the high wall.

She reversed until she was level with it. A bloke appeared from between cars on the carriageway, arms aloft.

'What the fuck do you think you're—?'

'Police.' That shut him up. 'What happened?'

'Car came haring along and smashed through those gates.'

Billy reversed another few metres then pulled the nose of her car onto the grass verge to illuminate the scene.

There was a gap in the fence that must once have been blocked by ten-foot-high metal gates that were scabbed with rust. One of them lolled on its hinges. The other had swung open. Chains – and presumably padlocks – lay on the ground. Kerry must have smashed right through them. Beyond the gates was a dirt-track service road, overgrown with weeds. It cut across the abandoned development site of Blythe Flats.

'I'm going to follow,' Billy called to the stranger. 'Close these gates after me in case anyone else has the daft idea to get off the motorway. It's not safe.'

'Right y'are.'

She accelerated over the grass verge and between the gates. Away in the distance, she spotted the shape of the minivan heading north-west across the wasteland. Billy set off again in pursuit. Full speed now, foot to the floor. The hire car bumped and rattled, but she kept it on the rough road.

It took her another half a mile to realise she wasn't scared, wasn't panicked, wasn't menopausally challenged. The only rush of hormones now was adrenaline.

And she had a fair idea where Kerry was headed; Howle Green. Her old home or the scene of her husband's accident. They knew they weren't getting away. Maybe they always knew. But they did it anyway. Murdered the man who had hurt their loved one and left him to die. They'd *gone all in* – as Dom might put it – with a hundred and ten per cent level of emotional commitment that Billy didn't really understand because she'd never gone that far herself. Except for once, with her daughter, and never did it again.

Billy scrabbled for her phone and eased off the speed. No need to force Kerry into an accident. She held down the button on her phone and gave Siri instructions to search for Howle Green on Google Maps.

'I found this on the internet,' said the silky voice.

Billy stuffed the phone into a cubby on the dashboard. It showed her moving through blackness towards a road marked ahead. Presumably, the track across the development site connected to this route. Kerry must have spotted the service road on the satellite map. Billy dodged an oil barrel and continued towards her destination. The GPS showed it was six minutes to Howle Green.

When Billy reached the road, there were more smashed gates. She had to knock one aside, wincing at the damage to the hire car. She drove along a lane and into the village, relieved to feel a proper surface under her wheels. It was desolate. And yet it wasn't late, almost half-nine on a Friday night.

Where was everyone? It was like a ghost town.

She gave Siri another instruction: call Dom.

It rang and when his voice came on the line it was tinny but welcome. She told him where she was heading, and he got on his computer to guide her towards Kerry's home.

'I'm coming up to the church,' said Billy. She heard the tippy-tap of his keyboard.

'Got the map on my screen,' said Dom. 'No ANPR, so I can't trace her vehicle. You must be on the B5602. Turn right after the church. It's about a mile to her address.'

She passed a dark pub.

Realisation hit her. The landlord. Could be the silver fox's pub. He said he wouldn't be back in time to open up. 'They come from the same village, the two women and Nigel Heathcote. They must know each other.' Billy cursed herself for being slow on the uptake, but then she hadn't known that the women came from Howle Green until half an hour ago. On the motorway, the women hadn't interacted with him at all. Why not?

You'd only avoid someone if you like them too much or too little.

'Why is Heathcote's pub closed on a Friday night? I know he's stuck in a traffic jam, but you'd think there would be other staff to open up . . .'

'Howle Green is divided by this bypass – three-quarters on one side and a quarter on the other. You're on the severed limb. RAF Howle is over the other side. I'm guessing the pubs on that side get more trade.'

She remembered from the news reports that Thomas Kingston had died on the bypass. It prevented traffic racing through
the village. But it also stopped any blood running through its
veins. Like so many villages that Billy had policed, this one had
no traffic, no business, no life. It was an animal dying of neglect.

'Here's the turn . . .' Billy nosed her way down a narrow lane
that became very rural, very quickly. Past a run of farm workers'
cottages, scrubby terraces with trampolines in the front gardens,
and cars on the road. One propped up on bricks. Then came
a dank bit under overgrown trees, until she pulled up outside
a nondescript bungalow. The house belonging to Kerry Wells
and, formerly, her husband Thomas Kingston. It was dark and
all the windows must have blinds or thick curtains because they
were blank. Her engine put itself to sleep, then tutted a couple of
times as it cooled.

The bungalow was squat, but the plot extensive. There seemed
to be outbuildings, probably land at the back. The sort of place
where people who aren't rolling in money might keep horses. It
was dark except for a floodlight around the side of a large double
garage. The minivan was nowhere to be seen. Nausea rose in
Billy's stomach. She'd lost them. She left the crime scene, then
lost the suspects.

'Anything?' said Dom.

She was about to admit her mistake when the floodlight by
the garage went out. It was on a timer. That meant someone
must have triggered it to come on. Someone *was* here.

'There are garages,' said Billy. 'I think the Toyota has been
hidden inside.'

'Wait where you are,' said Dom. 'Local police are on their way.'

Billy got out.

'Are you getting out?' Dom's voice hitched. 'I can hear you. Just stay put. These women are dangerous, they're suspected of killing a man. With a bloody bike spoke.'

A light came on inside the porch of the bungalow. The height of it suggested a torch or a phone.

'Hold up,' she said. 'Signs of life.'

The front door opened. Hyacinth stood there. Without her mother by her side, she looked her age. The body language was different. She was still tiny, but back on the motorway, she'd been hunched as though she wanted to make herself smaller, invisible even. Now she stood up straight, without her thumb in her mouth.

'Are you coming in?' Hyacinth called out.

'I'm waiting for the other officers to arrive,' Billy said.

'Don't you go in there,' said Dom's tinny voice, but Billy clicked down the volume.

'You have to come in.' Hyacinth glanced over her shoulder. 'She's going to hurt herself.'

Chapter Forty-One

9.27 p.m.

As soon as Billy stepped up to the bungalow, she realised why the windows looked blank. Boarded up. The house was ready to be demolished. A single wail from deep inside prompted Hyacinth to melt into the darkness, leaving the front door open.

Billy followed. She stopped in an empty hallway. Hyacinth stood in the mouth of a corridor that led to the rest of the house.

'She's going to do something,' Hyacinth said, her voice jagged. 'Not the first time either. That's why I made her live with me. I didn't want to come back here, it's no good for her. We hate this place.'

'What happened today?' Billy asked, keeping her voice level. 'With Lincoln Quick?'

'She's going to hurt herself, don't you care?' Hyacinth shook her head at Billy's priorities and disappeared into the gloom of the corridor.

Billy let go of the front door. It swung closed but didn't click. It plunged her into semi-darkness, the only light now a faint glow coming from along the corridor. She ran her hand across the wall, found a switch. It clicked uselessly. No electricity inside the house.

She followed Hyacinth, using the torch from her phone to light the way. With a free hand she patted her pockets, which contained the only other equipment she had in her possession;

the handcuffs she'd confiscated from Olly. She used her teeth to rip open the poo bag they were in and slipped the cuffs back into her pocket.

The whole house was devoid of furniture. Bare walls whose paint was marked with the ghostly pale squares left behind by long-gone picture frames as though the residents had decorated with a lot of very boring artwork. Strangely, the decor fit with what she knew about this family, especially Hyacinth who was blank, unknowable. And yet the floor was carpeted, plush underfoot, expensive. The bungalow held a faintly rank tinge of mud flats. It must have been empty and unheated too long. This house had been a home once. Now it was as cold and grey as grief.

'In here,' Hyacinth called out from a room that Billy discovered was the kitchen. Kerry bent over the sink, her phone lying face up on the counter, its cool torchlight illuminating the scene. Billy switched off hers and slipped the phone into her back pocket, the better to have her hands free. Kerry had her arm pressed flat on the edge of the sink so that her hand dangled into the bowl, wrist turned up to the ceiling. Her right hand. Her left held a thin metal rod. A bike spoke. From the stabbing grip, it looked like Kerry must be right-handed. If you really meant to kill yourself, wouldn't you use your best hand?

Hyacinth stood about a metre away from her mother, both palms held up.

'Mum, please.'

'It was me.' Kerry delivered the line over her shoulder.

'Mum!' hissed Hyacinth.

'My idea,' said Kerry, nodding at Billy. 'I forced Hyacinth, I threatened to kill myself unless she helped me. I killed him, I did the actual stabbing, she was only driving the getaway car.'

'I don't think so,' said Billy, keeping her voice low. 'Hyacinth is small enough to fit in the footwell. No disrespect, but you're not. And you were the one behind the wheel when we stopped, Kerry.'

'We swapped seats.' Kerry licked her lips. 'It was me in the back, I laid on the seat.'

'He would have seen you. It must have been her, hiding in the footwell.'

Kerry sagged. 'She's got her whole life ahead of her.'

'She's facing life in prison unless you tell me what really happened. Hyacinth is the one with the injury to her thumb from the stabbing.'

'We were supposed to get away!'

'But instead you got stuck in a traffic jam,' said Billy. 'Hyacinth is facing a murder charge, Kerry. You'll get accessory, three to ten years, but she'll do a minimum of twenty.'

'I'll kill myself.' Kerry pressed the sharpened point onto the thin skin of her wrist, making Hyacinth squeal.

'That's not going to change anything,' said Billy.

'It's not fair!' Kerry wailed. Her hand juddered and she must have broken the skin at last because she sucked in a breath of surprise. In the gloom, Billy could see a dark bloom of blood, but it wasn't enough to be coming from an artery.

'Kerry, why don't you put the weapon down while we talk?' Billy said. 'We can find a way to help Hyacinth. We can work this out.'

Kerry's eyes moved between her own wrist and her daughter.

Billy needed to keep her attention.

'Why did Lincoln Quick come back to England?'

Kerry breathed heavily through her nose, but the question seemed to steady her. 'Baby on the way. He'd been having it off with a girl in the village, she found out she was pregnant right after he left. She's due next week. I put two and two together.'

'Two and two don't add up to specific dates and times and flight numbers. How did you know when to expect him?'

'I'm going to do it,' Kerry hunkered down over her own wrist. 'I'm going to kill myself.'

'If you're dead, Kerry,' Billy softened her voice, 'Hyacinth will be the only one for the CPS to go after. This is a big story. It's going to make the news. You'll get sympathy because of what Lincoln Quick did to your husband. People hate him for skipping the country. But sympathy won't be enough to stop Hyacinth going to prison. You need to put down that weapon and help her.'

'How can I help her? It was all me, my fault, I was supposed to get us away from here, out of the country. If he could run away from justice, then so could we.'

That last point was only just landing, when Kerry turned from the sink and advanced on Billy. The black spoke dangled between her limp thumb and index finger. 'Look, I've confessed. Alright? It was me. I planned it, I did it. I got into the hire car at the airport while he was distracted by that bloke with the stupid beard. I laid down and waited until we virtually stopped and then I stabbed him in the neck. I used a bike spoke, that was a message, because he killed my husband with his bike. It was revenge for Tommy. And it was all my doing. Hyacinth didn't even know the plan. And now I'm threatening you. I'll kill you unless you stop me. Isn't that enough to arrest me?'

Kerry reached the place where Billy stood beside a large American fridge. The woman thrust her hands forward, weapon in her grasp. Hyacinth came up behind her mother, barely visible behind her shoulder, hiding in her shadow. Of course, it was Hyacinth who had used the bike spoke, jabbed it so hard her thumb got punctured.

The only sound came from Hyacinth's breathing.

Every sinew inside Billy tightened.

Her phone pinged and all the women jumped. In one movement, Billy whipped the handcuffs from her back pocket and knocked the spoke out of Kerry's hand. She snapped one cuff over Kerry's bloodied wrist and clipped the other half to the fridge door.

Kerry roared and tried to snatch her arm free. Instead, the fridge door swung open and hit her in the face.

Hyacinth lunged to pick up the spoke, but Billy was quicker. She snatched it up and held it aloft, showing that she would do no harm. Hyacinth backed away across the kitchen, where Billy noticed for the first time, there was a back door.

'Hyacinth, I need you to stay here,' Billy said. 'Just think, both of you.'

Kerry was breathing noisily through a nose streaming with blood. She tugged her hands a couple of times, but the handcuff and the fridge handle were firm enough to hold her. Her cut wrist must have been painful because she stopped fighting. She rolled her eyes towards Billy before a sob took her like a punch to the gut.

'He deserved it. He deserved to be punished.'

'I agree with you,' said Billy. 'Most people would agree with you. But murder is not justice.'

'There is no justice,' said Hyacinth from across the kitchen. She reached a hand behind her back and found the door handle. There was a metallic clunk of the lock and she pushed it open. Thin night air sidled into the room.

'Nobody cares about us. We're just victims. They don't even know our names, but they all remember his name. And everyone treats us like mugs. Everyone.' She directed this last word to her mother.

'Who treats you like a mug, Kerry?' said Billy. 'Who else is involved?'

'Shut up!' Kerry twisted around to face her daughter, gasping at the painful movement.

'Why are you protecting him?' Hyacinth moaned.

The mother said nothing and so the daughter stepped backwards out of the door, sinking into the darkness as though she had fallen off a cliff.

Chapter Forty-Two

9.33 p.m.

Outside the back of the bungalow, Billy stopped and listened for sounds of Hyacinth on the move. There was only the heavy breath of the wind as it molested the trees. The long moan of a car on the bypass. And then a soggy thump. She was over in the field, off to the left, must have fallen down.

Billy started across the garden in that direction, finding a tumbledown fence at the edge of the property that sagged enough for her to straddle the barbed wire and set off after Hyacinth.

Her knees jarred against unexpected mounds, her ankles bore the brunt of divots. Ahead, she could see a building. It was way off on the far side of the field, but she thought it must be the back of the pub. Heathcote's pub. Hadn't realised it was so close. That man again. Always there. Always watching. His pub had been closed when she passed it only a few minutes ago, so she had no idea what Hyacinth thought she'd find there. Maybe she was heading to the road. Or just running blind. Billy could make out the shape of the woman moving up ahead, a dark mass in the shadows thrown by trees.

'Hyacinth!' yelled Billy, but swallowed the end of the word as she fell hard and fast. She put down her hands and felt sand. A sandpit? At least it was a soft landing. She didn't bother to brush herself down, just got up and ploughed on until she reached another fence. Post-and-rail, easy to clamber over. Her feet landed on tarmac.

As she suspected, it was the rear car park of the pub. Billy approached the building. A cluster of picnic tables surrounded the back entrance for the smokers. The glass door was etched with a sign saying The King's Arms.

Should be The Silver Fox.

The handle dipped under her fingers and it swung open to let her into Heathcote's pub. A narrow corridor was lined with toilet doors. Signs saying Locker Room and Powder Room. A bend in the passageway led her into the bar. Billy stopped on the threshold.

Any room designed to be full of people was unsettling when empty. Too still. As though it was waiting.

'Hyacinth?' Billy's voice came out gin-thin.

She walked the length of the bar, feeling as though a ghostly barman watched her all the way while silently polishing a glass – as though bloody Heathcote was here in spirit. The silence put her senses onto high alert. The air was charged. Someone was here. A dining room stretched off to one side, all the tables empty. Chairs with backs too straight.

Beyond that was a smaller space, a snug. Billy startled slightly as she saw a figure, a woman, Hyacinth, sitting in the darkness.

'Are you okay?' Billy asked.

'Not really,' said Hyacinth.

Fair enough.

Billy found herself at the business end of the bar. She lifted the hatch and fumbled around for switches. She flicked a couple of buttons and wall lights illuminated, casting the room into a tobacco-coloured glow. Hyacinth put her hand over her eyes, although it was hardly bright. Billy sat down opposite her on a stool in the snug.

'Where're you running to?' Billy asked.

The girl shrugged.

'Just running?' Billy said.

Hyacinth nodded.

'What did your mum mean about going to another country? Was that your plan?'

'There was supposed to be money. We were going to get the ferry tonight, then flights tomorrow. We're all packed.'

'What went wrong? The traffic jam?'

'And the money. This is all we've got.' She put an envelope stuffed with cash on the table. And it was all fifties and twenties.

'That's quite a lot.'

'Less than ten grand. Not enough to get us where we were going.'

'Where were you going?'

'Tonga.'

Billy shifted in her seat. 'Tonga?'

'It looks warm.' Hyacinth picked at the plaster on her thumb. 'I was reading about this woman who ran the only cattery on the island, but she died in a tsunami, so I thought we could start a new cattery. Gap in the market, like. I volunteer at a rescue centre so I know what I'm doing.'

'Tell me about the money.'

'Mum's land.' Hyacinth raised her chin towards the back of the pub, and Billy realised she meant the field they'd just run over. 'Her inheritance. She sold it. But now it's gone.'

'You planned on using that money to get to Tonga?'

Hyacinth fixed her with a look of disdain. 'We should be able to buy Tonga. That land is worth millions.'

Billy tried to keep the incredulity off her face.

'It's got planning permission,' said Hyacinth.

'I see.'

'You think it's just a crappy old field, don't you? Well, it's not. It's part of an old golf course, fifty acres, planning permission for 200 houses and shops and work spaces. Regenerating the north side of the bypass. And it's all hers, no joke, we're not a joke.'

'I know you're not, far from it.' Under the table, Billy brushed sand off her thighs. Must have been a golf bunker she fell in. A building site for 200 houses could be worth – well, she had no idea, but it must be a lot. Even millions. Carl the estate agent would know, but she wasn't about to call him in for an estimate. It would certainly be enough money to live very comfortably in, say, Tonga. Of all the batshit crazy places to go on the lam.

Hyacinth looked up as though she'd heard a bell. 'I'm thirsty.'

After a beat, Billy got up. 'What do you fancy? Drinks are on you, with all that dosh.'

Hyacinth didn't smile. 'I'll have a J2O or something.'

Billy went through the hatch to the serving area behind the bar and took two bottles from the first fridge. Glass bottles. She put them back and found two plastic bottles instead. Wasn't about to arm the girl.

Back in the snug, Billy placed one in front of Hyacinth who didn't complain that she'd got fizzy water instead of a fruit drink. She took a mouthful and put it aside.

'She was having an affair,' she said.

'Who?' Billy tried to keep up with this conversational dog-leg.

'She told me today, in the car, on the motorway. She'd been having an affair when my dad died.'

Billy hummed.

Hyacinth snorted. 'Said she'd been planning to leave my dad.'

Kerry had been planning to leave her husband. There was money on the cards. And then the husband gets killed in a hit-and-run. Convenient. She thought of the CCTV video at the crossing. Lincoln Quick looking back down the road; that hesitation before deciding to leave the scene. A grimace or swearing or . . . acknowledgement? Could there have been a third person at the scene? Out of sight of the CCTV cameras? Giving instructions? If Kerry had been having an affair with Lincoln Quick, could she have incited him to kill her husband? Surely the original investigation would have checked her alibi?

'Where were you on the night your father died?'

'Am-dram. Village hall. *Murder On the Dance Floor.* Mum played a judge in a dance competition called Sherry Troll. And I played a stagehand who finds a dancer dead, crushed to death by a glitter ball.'

'Wow.'

Hyacinth shrugged modestly.

Billy's phoned buzzed, making the table vibrate. Message from Daisy – the list of Carl's sexual conquests and romantic battlefields. No time for that now, but Billy used it as cover to text Sergeant Gavin Glynn: Did K and H have alibi for night of husband death?

She kept the phone in her hand.

'Who was your mum having an affair with?'

'Wouldn't say.'

'Do you have any idea?'

Hyacinth shook her head.

'Was it Lincoln Quick?'

'God, no, do you think she's a cougar or something?' She chopped one hand through the air, knocking her water bottle over and watching the sparkling liquid fizz onto the carpet. 'She's over fifty, he's my age. Why would he even?'

Hyacinth probably did know the identity of the lover, even if she didn't realise it. Billy glanced around the snug, her attention drawn to a yellow T-shirt in a frame. Yellow jersey, in fact, from the Tour de France. Replica, obviously. Cycling again, what was it with bikes today?

'Why did you come in here, Hyacinth?'

'To hide from you.'

'How well do you know the landlord? Nigel Heathcote. He was on the motorway today.'

'Some. He's our neighbour. His daughter went to my school. She goes to the city every day, commutes, she's right up herself, works on the radio, only does the travel but acts like she's an A-lister, says she can't work behind the bar anymore on account of her shifts, but I think it's because of her fake nails, and she can't add up either.' The speech was the most Hyacinth had said in one go.

'What about her mother?'

'Died. She had a—' Hyacinth pointed vaguely to her chest so it could have been anything from breast cancer to a heart attack to a falling piano knocking the wind out of her. 'He didn't kill her.'

'Back at your house, you said your mum was protecting someone? Did you mean him, Nigel Heathcote?'

Hyacinth gave a deep-throated sigh of boredom.

'I know your mother didn't tell you who her lover was—'

'Lover!'

'Could it have been Heathcote?'

'I suppose it wasn't far to travel.' If Hyacinth intended the comment as a joke, she showed no sign of finding it funny. She picked the water bottle up off the floor, holding it with her fingertips while the last of the liquid dribbled over the table.

'Let's assume it was Heathcote,' said Billy. 'And let's assume that's who you meant when you asked your mum why she kept protecting him. Protect him how?'

'He's got the money.'

'What?'

'From the land. He helped mum to sell it. Brokered a deal. Now the money is in some kind of' – Hyacinth scrunched her face with the effort of finding the technical term – 'tax account. It's locked away.'

'Okay . . .'

'It's all online. Mum can't get it. He's got the login. And we were supposed to have cash and a card we could use, but we didn't get anything. Apart from this little bit that he gave me, which looks like a lot to you but is chips to us. We needed the money to go away.'

'To Tonga?'

'Right.'

'Heathcote is helping you? He knows?'

Hyacinth sucked her bottom lip deep inside her own mouth.

Fuck. She'd left Heathcote back on the motorway. With the crime scene. And all those people who didn't know what he was capable of. Why the hell had he put himself at the scene? At risk? The only reason she could imagine was that he'd made sure he was on hand in case he had to clean up a mess. He'd spent the whole afternoon watching, interfering, mansplaining – he wasn't the type to let two women do a job on their own.

'Stay here, Hyacinth. You do not move.' Billy got her phone and dialled Dom. He didn't answer after six rings, so Billy typed a message: I'm at pub, Kings Arms. Hyacinth here. Confession. Heathcote in on it. Pick him up

As she pressed send, an incoming text popped up from Gavin: Wife and daughter full alibi. Village event. Confirmed by 100 people

So Kerry couldn't have been at the scene on the bypass when her husband died. But what about Heathcote?

Billy texted back to Gavin: Wife and daughter killed Quick. Wife having affair with Heathcote, pub landlord. Did he have alibi?

She watched three dots scroll for a moment before Gavin replied: I'll check, I'm on my way to you

Billy: Bring in Heathcote

Gavin: k

Billy went back to the snug and looked at the yellow jersey.

'Is Heathcote into cycling?' Billy asked.

'His tight shorts make me gag.'

'He cycles?'

'He was in a racing team with Lincoln Quick.'

He'd said he didn't know Quick. No, Billy corrected herself, he'd said he didn't know Chad McClusky.

Fuck.

'Did they cycle together?' Billy asked.

'The whole village thinks he's partly responsible for my dad, because the cyclists used to go around too fast, like if they hadn't always been speeding, the accident might not have happened.'

'People believe it was just an accident?'

'No, people believe Dad could've been saved if someone had called an ambulance so it was kind of a murder.'

Not legally, but Billy took her point.

Her phone buzzed. She glanced down.

Glynn: Heathcote alibi for night of Kingston's death. Dinner with friend

That sounded familiar.

Billy wrote: Who was the friend?

Glynn: Name of Tubby Eden

Same friend he met at the Wharf today. Two deaths, two alibis from the same man. Billy found Eden's number in her notes. She could smell resolution in the air like the yeasty hangover of the night before.

Chapter Forty-Three

9.46 p.m.

'What happened to the spoke?' Hyacinth's voice was in her boots. 'The one Mum had?'

The deadpan delivery gave Billy a chill.

'The bicycle spoke? I left it behind.' Billy realised she must have dropped it in the field when she fell. Lucky she hadn't landed on the thing. Lucky also that she hadn't brought it in here with this woman who killed a man earlier in the day. Billy had no more cuffs, nothing with which to restrain Hyacinth. The woman sat quite calmly, but there was a bridled energy about her, like the invisible force around an electric fence.

'My mum could get hold of it and hurt herself.'

'She can't get it, she's fine, she's out of trouble. Stay there, please, I have to make a call.'

Billy dialled the number of Tubby Eden, Heathcote's alibi, but it rang out. There wasn't even a voicemail. She thought for a moment, cautious as to how to word her next question to Hyacinth.

'Did Heathcote ever talk about the night your father died? As a pub landlord, he must hear local gossip.'

'Course. They talk about it all the time, the whole village, and him and Mum.'

'What does Heathcote say happened that night?'

Hyacinth squeezed her plastic bottle and it crackled. 'He says Dad didn't stand a chance. He didn't do nothing wrong, he was minding his own business, even looked before he walked, but he couldn't have known someone was coming the wrong way down the road. So it was just bad luck.'

'Did the police show you any CCTV?'

'No. I wouldn't want to—'

'Has your mum seen it?'

'Don't think so.'

'Then I wonder how Heathcote knew that your father looked before he walked?'

Silence from Hyacinth.

Billy's phone rang. Tubby Eden. When Billy picked up, Heathcote's friend sounded put out to hear from her again.

'Last year, Mr Heathcote was questioned by police after a man called Thomas Kingston died.'

'Terrible thing what the American did, skipping the country.'

'I have a question about Mr Heathcote's movements that night.'

'I told the police everything I knew.'

'You covered for him?' Billy said.

'Yes, well, no. What do you mean by "covered for him"? I explained that we had met for dinner.'

'And did you? Meet for dinner?'

'Well, yes, yes, we had done.'

'You *had* done?'

'We had met earlier that day, yes.'

'But the incident happened late in the evening. Were you with Mr Heathcote the whole time?'

'This is awkward . . .' The line went quiet.

'Did you lie, Mr Eden? Can I remind you that providing a false alibi is an offence?'

'It wasn't a false alibi, just a bit of leeway.'

'What sort of leeway?'

A hard breath rattled the line. 'Nigel was having a fling. With a married woman. He met her after our lunch. That's who he was with when the accident happened. But he couldn't tell the police because it might have exposed her. Her husband was violent, do you see? He was doing it to protect this lady.'

Eye roll.

'It was perfectly legit, consensual and all that, just not something he could tell the police or they might have gone after this married woman to confirm the alibi and that would have been awful for her. So I just—' He stopped.

'Lied?'

'Stretched the truth. It didn't matter, did it? It was clear that this American chap was the one who ran that man down and killed him. Then he skipped the country. Guilty as sin, obviously, or he wouldn't have done that. So it was just a white lie, in the end, to protect the woman as much as Nigel.'

'Did you meet this woman?'

'No.'

'So you only have Heathcote's word that he was meeting someone after your lunch?'

A beat, long enough for realisation to drop. 'I suppose I did, yes.'

Billy sighed. 'Was today a white lie as well?'

'No, absolutely not. Today we were at the Wharf the whole time. You can check. Nigel paid with his credit card.'

'I've already confirmed that. He didn't leave early and give the credit card to you or anything like that? That would be perverting the course of justice. I'll be checking the CCTV later.'

'No, no, we were both there, together. I swear it. Mother's grave.'

Billy found a polite way to tell him he was a dickhead and rang off.

Her phone pinged with a message.

Dom: On my way. Ten mins. Hold tight

He must have driven like the wind. But Billy felt a warm hit of relief.

She needed to talk to him about motive.

Love, lust, loot and loathing.

The classic motives for murder.

Of those four, Kerry and Hyacinth had been motivated by loathing.

Loathing meant revenge.

Billy twisted open her plastic bottle of water and took a swig.

That bothered her. She took Dom's point about *going all in* – perhaps Kerry and Hyacinth had been so committed to their husband and dad that they didn't care about the consequences – but murder usually pivots on how much there is to win or lose. With revenge, there is very little to win and, potentially, a lot to lose. There is no gain in revenge, really, except satisfaction. Do people kill for satisfaction? They might feel vengeful, even murderous, but do they actually go through with it? Not often. Which is why revenge is the least common motive for murder.

So, Kerry and Hyacinth, murdering a man to avenge their husband and father?

Even if they were going all in, emotionally . . . ?

Avenging a father was a bit . . . well, a bit Batman.

It didn't feel right. Despite the plaster.

And there were too many questions.

Who gave them Lincoln Quick's flight number?

How did they know the CCTV was switched off?

Who blackmailed the estate agent?

Carl Doors said that a man had called him. Someone who knew one of the women. Or perhaps someone from the hotel where they met . . .

Billy looked around the pub. On the wall behind her, one stencilled sign pointed towards toilets, another sign pointed to rooms.

This pub had hotel rooms.

Billy got up and went to the hatch in the bar. There was an old desktop computer and a pile of paperwork. Several large hardbacked notebooks. Billy opened one and found a list of numbers for suppliers and contractors. She opened another and found restaurant bookings. The third was hotel reservations.

She flipped back page by page. There were only four rooms above the pub, so it wasn't a laborious task. She went back almost three months before she saw the name Carl. No surname. But there was a mobile number. She checked it against Doors's details in her notes: it was his phone, his real one, not the burner.

Carl had stayed at Heathcote's pub. Presumably with one of the Dizcreet women.

Then she remembered the text message from Daisy.

Carl's sex pest list. She could've saved herself the trawl through the reservation book if she'd checked that first . . .

A proper detective would do the right things in the right order.

She opened the message. Read the details. There were five women and five hotels.

In fact, he'd met three of them here. Nice touch, Carl. Classy.

She photographed the page and texted it to Dom. They would have to check this out, but she surmised that Heathcote had found out why the stranger was visiting his pub with different women, then blackmailed Carl and got him to do the dirty work at the airport.

Again, if it hadn't been for the gridlocked motorway, the estate agent would have sailed off home. No doubt he would have heard about the murder afterwards, but he was hardly likely to come forward and confess his part in it, was he?

It made sense that Kerry and Hyacinth had had help. The extra push they needed to go through with it. They were motivated by loathing, Heathcote was motivated by . . . what? Loot, must be. The money from the sale of the land. Billy got out her phone and texted Sergeant Gavin Glynn, asking if he'd arrested Heathcote yet.

The answer came back at once: He's gone

Heathcote had done a runner too.

At that same moment, Billy heard a crunch of tyres in the car park and a beam of light scuttled around the walls.

Please be a patrol car.

Billy staggered under a blow from behind. She put her hands up to defend herself, but it was Hyacinth rushing past, shoulder-barging her as she ran to the hatch in the bar.

'He's coming!' She sank out of sight behind the counter.

Billy messaged Dom as fast as her thumbs could move: Heathcote pub now

The phone pinged in reply, but Billy didn't read the message because the front door of the pub rattled, a key jangled in the lock, and a moment later Heathcote batted the swing doors out of his way. They thumped back against the walls, juddering. He was up on his toes, pumped, a prize fighter. Arms wide of his body. Quite a contrast to the studiously laid-back publican from the motorway. No longer silver fox. More silverback.

Billy resisted the urge to run behind the bar and join Hyacinth.

She stood her ground as Nigel Heathcote narrowed his eyes, strode across the pub and jostled her backwards into the snug.

Hyacinth aka The Daughter

Honda Zoomer 50cc

Glass tumblers line up under the bar. A miniature army ready to quench your thirst. Their surfaces reflect the colours of the optics.

I did a shift behind this bar. He tried to make out like I was doing him a favour even though I worked ten hours with him breathing down my neck, saying '*Not like that*' and '*Do it like this*'. He spent so long on all the things that could go wrong with the optics that I couldn't think straight after. You have to press the tumbler hard against the plastic lever to make the liquid come out. But the edge of the glass is too thin. I could almost feel it shattering, slicing my skin, hot blood down my wrist, shards spiking my face, the bottle jumping free of its bindings and smashing on the floor, liquid round my ankles, the whole pub watching and thinking I must be stupid 'cause he'd warned me enough times. He put the image into my head like a gif that wouldn't stop repeating. In the end, Nigel had to work the optics himself because he'd been such a control freak about it. Only had himself to blame.

The front door of the pub crashes open.

The tumblers shiver.

Here he is now.

Heavy footsteps and then the woman, the policewoman, says '*You're under arrest*' and '*Stop right there*' and he says, '*You shouldn't listen to the girl Hyacinth because she's thick as white bread,*' and that's a shocker because I wasn't too thick to work in his pub, was I?

Then he says '*That Kerry Wells*' (since when does he use Mum's full name like that?) '*That Kerry Wells ought to be ashamed for getting her own daughter caught up in this mess when the girl has her whole life ahead of her,*' and the policewoman is trying to get a word in edgeways but he keeps on '*If her mum goes to prison there's no one to look after her,*' and that's rich because I've mostly had to look after myself since Dad died. Got a little flat off the council and even had her sleeping on the sofa. So who's looking after who, then? But the policewoman gets fed up of his mansplaining and interrupts. She goes back to him being under arrest and then all of a sudden it goes quiet.

There's shuffling.

Feet.

Definitely. Feet.

And breaths.

Maybe he stopped the policewoman from talking because he doesn't want to be arrested, which is fair enough but you can't stop an arrest once it's started, it's like a wave, or an oil tanker, something like that, an avalanche.

The glass tumblers are still. Nothing moves.

Except the shuffling.

The odd huff.

Today has been about justice. Nigel said it was closure and Mum said '*Yeah, that's what we need, closure.*' Justice served cold. But then in the car on the motorway she said it had been a trick, Nigel is after the money. He thought he could shag his way to our money by marrying Mum (gross), then he saw a chance to get the money without having to shag her (fair enough) and Mum worked it out while sitting there on the motorway.

There's a little shelf above the tumblers with a corkscrew on it. The tip feels cold and sharp against the pad of my pointing finger.

Before, when we were talking, the policewoman asked how Nigel knew that Dad had looked both ways before he crossed the road.

How did Nigel know that?

Unless he was there.

How many times has he said that knocking someone over might have been an accident, you could forgive that, but the one who stood by while Dad died in the gutter was a real heartless bastard.

But Nigel must've stood by while Dad died. And then he let Lincoln Quick take the blame. And today he stood by while Lincoln Quick died. And he let us take the blame.

On my knees, I can peer over the bar. He's got something around the policewoman's throat. She's kicking her feet, dancing on tiptoes.

That isn't justice. Or closure. She hasn't done anything to us.

Can't watch.

But even when I shrink down behind the bar, the sound of her spluttering and gasping follows me. It makes me think of a pug we had in the rescue centre. Those bulging eyes, that panicky expression when they can't breathe, *I'm slowly drowning here*.

In my hand, the metal levers of the corkscrew rise like the arms of a surrendering man.

Don't like Nigel anymore.

I realise I'm on my feet.

And I'm not doing his dirty work anymore.

Chapter Forty-Four

9.53 p.m.

Billy found a surprising level of clarity while being throttled with an elastic bungee cord. The kind you'd use to fix something to the back of a bike.

Flashing lights mean lack of oxygen.

The exploding feeling comes from pressure in the brain.

I should be unconscious by now.

He wasn't doing it properly. He was failing to kill her.

That annoyed her. She wasn't going to be beaten by some amateur.

She was trained for this. She knew what to do.

She pressed her hands together under her breasts like she was praying, then thrust them up between his arms, blasting his wrists aside with her elbows. He staggered and released the bungee cord enough for her to swing her right arm and slam the heel of her hand up under his nose. He grunted and folded over double.

Billy was about to follow it up with a knee to Heathcote's face, just to make sure, but behind him, Hyacinth stood stock-still with one arm raised. Something glinted in her hand, silver in the sepia air. It wasn't a bike spoke; it was too thick. She made eye contact with Billy for a long second, then focused her attention on Heathcote's back. Hyacinth took a step forward and plunged the shiny thing into his spine.

He sucked in a deep breath.

'Get off her!' Hyacinth shouted at Heathcote, even though he'd let Billy go a solid ten seconds earlier.

He roared as he toppled sideways onto the sticky carpet.

With Heathcote down for the count, Billy fell back onto a banquette, sucking air and massaging her burning throat. It felt like he'd cut it open, but there was no blood when she looked at her palms. It hurt more now that he'd stopped. Hyacinth fled behind the bar again. Billy let her go, she wasn't in any state to take the girl on too.

The landlord crawled around the snug like a wounded bear, trying to grab the object stuck in his mid-back.

Oh, it's a corkscrew.

Its metal arms flapped in mockery of his attempts to reach it.

Bit of a late tackle from Hyacinth. But Billy was just grateful she didn't have to fight Heathcote again.

Fuck his kidneys.

The corkscrew was a bit higher than that.

Fuck his lungs.

She picked up the elastic bungee cord and shoved Heathcote towards the bar. She got him down into the recovery position – putting him on his back seemed unnecessarily vindictive – and tied his wrists to the brass foot rail that ran along the bottom of the bar.

With him secured, Billy flopped back and sat on her bum.

Her throat was burning.

And Hyacinth had gone.

Ffs.

As Billy got out her phone to text Dom, she heard a thunking din. Sounded like a helicopter. So that's how he'd got here so quickly.

She rolled onto her hands and knees, pushed herself up to her feet and gave her back a moment to straighten itself out. Could be worse, could have a corkscrew in it.

By the time she walked back along the bar and into the car park, the police chopper had landed in the million-dollar field.

A man in a suit came her way, loping over the tufted grass as though he had one leg shorter than the other. He was silhouetted against the backdrop of helicopter lights, his jacket pulled out like a cape by the eddy of the blades, and she noted that he'd grown leaner from all that thinking about stand-up paddling, and she also noted that she'd just referred to him as wearing a cape and that shit needed to stop right now because if he heard her say that out loud she'd never hear the end of it, and as he reached the car park and climbed the fence into the light, he slowed at the sight of her and sagged a little in relief.

Superintendent Dominic Day slapped his right hand on his heart.

Billy waved.

He wore horn-rimmed glasses and morning-after-the-night-before stubble. Billy thought she should send instructions to Carl – *Mate, this is how you do facial hair.*

Chapter Forty-Five

9.59 p.m.

Billy could smell herself. Ripe skunky pheromones. She must have pumped out some bestial defence mechanism while Heathcote was throttling her. She hoped Dom didn't notice it. He'd bundled her around the side of the pub out of sight of everyone else and grappled his arms around her.

Pinned to the wall of a pub for the second time in ten minutes.

But this time was better.

She crossed her arms over his back, pressed her chin onto the shoulder of his blue suit, staring down the slope of skin that swooped into his shirt.

There was no getting away from it, he must be able to smell her.

His hands slid from her lower back up her sides to cup her shoulder blades. He leant back so they were face to face. Her hair caught in his stubble and formed a cobweb between them. Someone called her name from the car park. He swore. Their hands slid to each other's elbows. He tugged her forward and kissed her between the eyebrows, so hard she felt his teeth under his lips.

She walked away giddy to the ankles.

Must be the jet lag.

After that, various killers were located and detained.

Then Billy and Dom sat on high stools at the bar, nursing bottles of beer, and cheese and onion crisps that he'd requisitioned from behind the counter. The food hurt her bruised throat, but she was starving so she soaked each crisp in beer until it went soggy enough to swallow.

'Just when I thought we could make a life together,' muttered Dom.

'Get salt 'n' vinegar next time.'

The air between them was ripe. And it wasn't just her skunky smell. It was the way it feels when you walk into a house where someone is cooking up a storm. Like you want to hang up your coat and take a seat at their table.

'Thank God that dozy woman was here with her corkscrew,' said Dom. He side-eyed Billy for a reaction.

She only hummed. No need to clarify the exact timing of events. Hyacinth's late tackle. What was a couple of seconds between friends? Let everyone think it was self-defence, she'd saved a police officer from a nasty man. Maybe the judge would take it into consideration when sentencing. Hyacinth would be going to prison. The only question was for how long. Her mother too.

'You know, Kerry Wells tried to give me a blanket when we were on the motorway,' said Billy. 'She wanted me to cover the body so they didn't have to see it, said it upset Hyacinth.'

'She was thinking about forensics.'

'Would have explained why Hyacinth's DNA was all over the car.'

'Not daft,' said Dom.

'Heathcote is worse,' Billy said. 'Got someone to do his dirty work twice. Quick killed Kingston. Hyacinth killed Quick. Seems unfair that Heathcote will get off with less.'

'Accessory to two killings, if the judge gives consecutive sentences, could still mean twenty.'

Billy opened a second packet of crisps while she watched the goings-on outside the pub. They'd pinned the double doors aside, SOCOs lugging boxes of evidence from the flat above the pub.

Paramedics were treating Heathcote who lay on a gurney, corkscrew lodged in his spine. They strapped him in, got him settled, loaded him into the back of the ambulance. It pulled away with blue lights but no sirens. Not necessary on the country lanes.

A patrol car pulled into its place. A cadaverous-looking black guy wearing civvies got out of the passenger seat.

'Gavin Glynn?' Billy asked, getting to her feet. Dom snapped the crisp packet taut and tipped its dust into his mouth before following.

Sergeant Gavin Glynn was apprising Billy of the operation on the motorway.

'Charlie Boy would only tell me what he saw if I agreed to drop the drug charges. He asked for immunity. Thinks he's in *Narcos*.'

'And what vital piece of evidence did young Mr Neale leverage to make this brilliant deal?' she asked.

'He reckons he saw our man Heathcote hand something to the girl.'

'Hyacinth? Did he see what it was?'

'Nope. He said you'd gone down the carriageway to check on the car bomb that turned out to be a false alarm, and Heathcote watched you go then walked across to the minivan and dropped something into her hands. Didn't say nothing, just did it quick while no one was looking and got back in his own car. Didn't see Charlie Boy in the trees. He said he tried to tell you what he saw, but you wouldn't let him.'

'He kept trying to make a deal. I thought he was making it up.'

'He might be,' said Gavin. 'Or it might have been that envelope of cash she had. Forensics will tell us the answer to that little mystery.'

Billy hummed her agreement.

'So does Charlie Neale think he struck a bargain with you for immunity, Sergeant Glynn?' she asked.

Gavin grinned and scratched his bald head. 'He does.'

'And did you tell him that there are no drug charges 'cause all the evidence would be inadmissible?'

Gavin winked. 'Nah, if he can withhold, then so can I.'

They stepped aside as another vehicle came steaming into the pub car park. Hatchback, sensible, teal with matching trims. Very girly.

A woman got out, one hand supporting a pregnant belly that Billy could see had dropped. Must be due soon. Behind her back, out of sight of the menfolk, Billy crossed her fingers and glanced up at the sky to make the mental plea that had become her habit at the first sight of any new mother.

I wish better for you.

But as Billy went to intercept the woman in the teal car, a high whine sounded in her ears. Tiredness. Jet lag. A warning sign. It sounded like a mosquito that is going to keep bugging you until you get up and squash it.

'Can I help you?' Billy said.

'I live here. What's going on? I got a message from my father.'

'Nigel Heathcote?'

'Yeah. I'm Amy Heathcote.' Her tone contained a hint of *And what of it?* that told Billy that the arrogance doesn't fall far from the tree. Name rang a bell, but she couldn't place it. Amy Heathcote was all fake blonde, fake eyelashes, fake fingernails.

She swung a set of car keys around her index finger and caught them in her palm. The keyring advertised the local radio station. And then Billy had it.

'You read the travel news on the radio.'

Amy wobbled her head in surprise. 'I don't get recognised all that often!' She smiled now, stopped spinning her keys, flattered and disarmed and willing to spend time on this person who wanted to admire her.

'I was stuck on the motorway earlier,' Billy said, 'I heard your name every hour.'

'Every hour on the hour!' said Amy in her radio voice.

'Oh, you do that very well,' laughed Billy.

'Well, I get paid to, so.'

'When are you due?' asked Billy.

'Next week. Today was my last day at work. Where's my dad? Why are the police here?' She twisted a ring on her finger. A shiny wedding band. Just like the one the dead man was wearing.

'Do you know a Lincoln Quick?'

Amy said nothing. Stopped twisting the ring and put both hands on her belly as though she was blocking the baby's ears.

'I have some bad news, I'm afraid.' Billy said. 'Lincoln Quick was killed today. An incident on the motorway.'

'What?' Amy steadied herself on the open door of her car. But her eyes flitted to an upstairs window and Billy wondered how much she knew about her father's whereabouts that day.

Billy signalled to one of the PCs, told him to fetch a chair. She didn't want Amy going into early labour. Dom came over, a moth to the flame of any drama.

'Superintendent Day, this is Amy Heathcote, Mr Heathcote's daughter. I'm afraid I've just had to inform her that the father of her unborn child died this morning.'

Dom frowned at Billy but managed to offer his condolences.

'Her husband, Lincoln Quick, came back to England for the birth,' Billy explained.

'He wasn't my husband, it wasn't legal yet. He sent me the ring in the post.'

'What were you planning to do after the birth, Amy? He couldn't have stayed in England, there's a warrant for his arrest, so were you all going to move to the States?'

'It was an idea . . .'

'Your father didn't want you to go?'

'This is his grandchild.' She rubbed her belly. 'Of course he didn't want us to go.'

'How did he feel about you and Lincoln Quick? After all, the man killed someone.'

'It was an accident.' The chair arrived and she lowered herself into it. The effort seemed to take the fight out of her. 'They were friends once, but my dad found it hard to forgive after the accident.'

'Hard enough to have Lincoln Quick killed?'

'What? No! My dad is many things, but—' She stopped, blinking rapidly, her face working hard to stay straight. Her false eyelashes looked like moths flinging themselves against a windowpane, mad for the moon. After a few seconds, they exhausted themselves. She stared down at her swollen belly. 'I can't imagine him doing that,' she said. Which meant that perhaps she could.

'Presumably you knew Mr Quick's flight details today?' Billy asked.

'Yeah.'

'And your dad knew too?' Billy said.

'Yeah.'

Dom pushed his fists deep into his trouser pockets and rocked onto his heels. This was his equivalent of an Andy Murray fist pump, a Ronaldo siu, a Usain bolt. Only in a manner that befits a senior officer of the law. But his celebration had gone off too soon because Billy wasn't finished. She still had the high whine in her head.

'Amy Heathcote works on the radio,' Billy explained to Dom. 'Traffic and travel. Do you have access to a feed from surveillance cameras on the motorway? To monitor traffic?'

'Yeah, why?'

'And a work laptop, maybe, that you bring home?'

'Yeah.'

'Is it possible that your father might have accessed the CCTV feed to discover where it might have been switched off?'

Amy held her stomach.

Billy didn't press the pregnant woman any further. No need. But Dom gave her a very significant look over the top of his new horn-rimmed glasses that suggested he would get her salt and vinegar crisps next time, or any bloody flavour she wanted.

Back on the motorway, Billy got out of the patrol car and stood on the hard shoulder. A message on the gantry said Obstacles on Road. Indeed there were. A white tent around the sedan, and a body bag waiting for its grim cargo.

Diplomatic immunity had protected Lincoln Quick from the law, but not from two unhappy families who caught up with him quicker than justice.

On the other side, the abandoned 'CIA' van had been towed and the traffic was moving. A speed limit remained in place, drivers rubbernecking as they passed, but the steady flow gave Billy a pleasing sensation of flushing the drain. She wondered if the guy who had suffered a panic attack was alright. And the pound shop angels, the old man who broke into the van on her behalf, Finger Food and her looted canapés . . .

'I'm impressed you kept this lot under control,' said Dom.

He kept checking on her with sideways glances. Taking her in in sips.

Most of the vehicles comprising the grid had moved from the carriageway. But until the body of Lincoln Quick was removed, only the outside lane could reopen. Cars sneaked past behind a fringe of orange cones.

The estate agent's vehicle had been towed by forensics. Carl Doors taken to the nick for questioning. Likewise Charlie Neale with the slack bladder and his sister's Renault. Further back down the road, the abandoned vehicle had been investigated and found to be free of car bombs, as she suspected, so it had been hauled away. Only vehicles belonging to witnesses were left.

Daisy Finch and Olly Sims were giving statements to separate officers.

'Those are the two with the handcuffs,' said Billy.

'Fluffy ones?'

'Just plain. I used one set on the drug dealer.'

'Still got the other set?' Dom asked.

'Behave. I used the other set on Kerry.'

As they watched, an officer handed his device to Daisy and asked her to sign a statement. She did it with the awkward flourish of someone using another person's tablet. The constable thanked her with a nod. Without a backward glance, she walked to her hot-pink Nissan Micra and started the engine. A moment later, she mirror-signal-manoeuvred into the line of traffic. Olly, still occupied with his officer, shot glances over his shoulder as Daisy left him behind.

'Aye-aye,' said Dom. 'Looks like she's given Handcuffs the heave-ho.'

Mandeep Singh took Daisy's place to give his statement. The boy, Jack the poker shark, had already gone. Billy caught the sleeve of a passing constable, who was leading a dog on a string. She recognised the sandy mutt who'd got loose in the traffic jam.

'Was the kid alright?' she asked.

Jack had gone in a panda car to be reunited with his mum.

'Where did this dog come from?' the PC asked Billy, looking put out. 'Didn't you find his owner?'

'I was busy with a few other matters,' she said.

'He's a right doofus. Peed in one of the forensic bags. I'm thinking someone took the opportunity to dump him.'

'Behave,' said Dom. 'Someone'll be frantic further up the carriageway. Get going.'

The PC muttered as Dom ushered him and the dog against the flow of traffic to find its owner.

'You know when you go on holiday,' said Billy, 'and you get chatting to a stranger and then spend all your time with them and feel like you're best mates?'

'And then you come home and never see them again?' said Dom.

'That's how I feel. This wasn't a case. It was more like a hostage situation.'

'You've been through an ordeal.' Dom pressed his arm against hers, and she pushed into his side for a brief moment.

Pat Mackey and Charlotte McVie stood with a female officer between them.

'But I don't want it,' Charlotte was saying.

'You can't abandon a vehicle on the motorway,' the police-woman insisted, shaking her braids.

'I only came back because the traffic was no better on the side roads. I was never going to get a taxi so I thought I might need the car after all, but Pat has offered to drive us, so I don't want it now.'

The policewoman widened her eyes at Billy in disbelief. 'It's worth about eighty grand.'

'All you care about is the value of property. It's his, let him sort it out.' Charlotte turned to Billy as she approached. 'I never want to see him again. I never want to see his car again. I was planning to push it in the reservoir to piss him off, but I can't be bothered. I want to get my sons and get on with my life. So scrap it, if you like. Or keep it. Have it for free. Have it towed at his expense, I don't care.'

The young officer started to protest, but Billy stopped her with a finger pressed on the arm.

'I'm sorry you haven't found the police more supportive, Charlotte. From what I've heard, we let you down.'

Pat said some things that reminded Billy of why she'd once named her Foul Mouth. She tugged at one of Charlotte's long sleeves. 'Have you seen the state of her? Show her, love.'

Reluctantly, Charlotte pulled up one sleeve to show her wrist.

It looked like a skillet, her skin bruised into stripes from black through to lemon.

'He used cable ties,' said Pat.

'Charlotte—' said Billy, but the woman held up one hand.

'He does it when the boys are away. But it won't be long before he turns on them. I've got to get them away from him.'

'You should report this.'

Charlotte turned her shoulders so that she faced Billy full on. 'Fuck you,' she said, calmly. 'One time, you lot sent two officers to my house. At last, I thought, at bloody last. But they were two young lads, probationers, more interested in his Tesla. They looked all around the house as if, well, as if a few bruises isn't a bad price to pay. They took notes and said there was nothing they could do. My word is not enough. My bruises aren't enough. So don't talk to me about reporting it. I paid for a solicitor and got a non-molestation order, so I'm dealing with it myself, thank you very much.'

'I'm going to help,' said Pat. 'We're going to collect the boys and I'll drive them to the refuge. It's not far from where I live. I'm going to keep an eye on them, do some babysitting, help house-hunt and what have you.'

'You're kind,' said Charlotte.

'Not particularly. It'll be good for me too,' said Pat. 'Keep me out of trouble.'

'The parking ticket has been wiped,' said Billy. 'My colleague here took care of it. There's no trace of it on the system. Your husband won't be able to find you.'

Charlotte nodded a curt thanks. She turned on her heel and went to gather her belongings that lay on the grass verge.

'Look after her, Pat,' said Billy to the nurse. 'And look after yourself.'

Billy surprised herself by giving the woman a hug.

'Best day out I've had in bloody ages,' whispered Pat in her ear. 'By the way, you smell rank, love.'

Pat went off after Charlotte and they walked together to her Volkswagen.

'You hungry?' Dom appeared by her side.

'Starving,' said Billy. 'But I've got to find my suitcase, I used it to hold up the cordon. And my gin!'

'You're a right Girl Scout. There's a good pub not far from here, on the canalside.'

'Not the Wharf?'

'What's wrong with the—?'

'I don't want to run into Tubby bloody Eden. There's a curry house near my place. I've been craving a proper Balti for six months. I'll have a shower while you pick it up.'

'Roger that. You could do with a shower.'

Tired as she was, she'd have to drive as he wasn't insured on the hire car. They located her case and, miraculously, the bottle of duty-free gin that must have narrowly escaped being run over by Kerry Wells in the minivan. As she drove away from the crime scene, the body bag containing Lincoln Quick was being pushed inside the ambulance on a gurney.

'What a mess that man left behind,' said Dom. 'So many lives could have been saved if he'd done the right thing.'

'In a funny sort of way, it was touching to see a community come together like that.'

'What, to kill a man?'

'Mm-hmm.'

'Good to have you back, Sergeant Billy the Kidd.'

She put her foot down as the motorway opened out to three lanes. Ahead, the gantry sign flashed green arrows pointing into the night. To the reassuring thrum of tyres on tarmac, Billy headed home.

Acknowledgements

First thanks goes to all the readers, reviewers and book bloggers for getting behind the wheel of *Dead Mile*. You've reached the final destination on a long trip that started when I was stuck in traffic on the A23 near Gatwick, and thought how *interesting* it would be if the chap in the car next door slumped over the wheel with a knife in his back. And so, a story was born. I do hope the chap got home safely; he did nothing to deserve this.

Around that time, I formed a new writers group. Tentative early chapters of this story were shaped by founder Hovarians Esme Hall, Julian Morgan, Jezza Donovan, Stephie Woolven, Jane Crittenden and Diane Stephenson. Thanks to the entire Hove gang for their Monday-night company and insights.

I raise my pincers to Kate Simants, Louisa Scarr-Sam Holland-de Lange, Niki MacKay, Clare Empson, Heather Critchlow, Fliss Chester and Rachel Blok for their cheerleading. Also Dom Nolan, T. E. 'Tim' Kinsey, Rob Scragg, Adam Southward, Simon Masters, Susie Lynes, Harriet Tyce, Liz Mundy, Victoria Selman, Elle Croft and Phoebe Morgan have been generous with advice and antics. Kudos to James Delargy for noting that *Dead Mile* should be a novel and not a short story. Cheers to my Brighton buddy Jack Jordan for getting me out of the house now and then. Also to Graham 'gbpoliceadvisor' Bartlett for his expertise, some of which Billy and I wilfully ignored which is why any procedural mistakes are my fault.

This book would not be in the world without Jordan Lees at the Blair Partnership. I'm grateful for his enthusiasm and for knowing that a writer cannot be trusted to react normally to news, even good news, without careful preparation. Huge thanks to Georgie Mellor and the foreign rights team who made a writer's dream of foreign titles come true.

I'm so fortunate to have Kelly Smith at Bonnier Books as the driving force behind *Dead Mile*, and relieved that her fellow editors Laura Gerrard put the brakes on my over-extended car metaphors.

Greetings and gratitude to Danielle Egan-Miller at Browne and Miller in Chicago for the kind support and expertise over the past several years.

Thank you, Mark, for the tea and sammiches. I wouldn't want to be stuck in a jam with anyone else. And Lydia and Frank – your company and playlists brighten my way.

Enjoyed *Dead Mile?*

Read on for an exclusive preview of Jo Furniss'
next thriller

GUILT TRIP

It is 3 p.m. on Monday. Emily Smith is summoned to her
daughter's school, an elite hothouse where pupils work
under exacting standards of grades and behaviour.
She is reassured that Olivia isn't in trouble, but the
news is even worse – her child is missing.

A minibus carrying the swim team never arrived back
after a meet. The police don't know where it has gone
and the school's 'no phones' policy means none of
the kids have a device. They have vanished.

As time ticks by and fear becomes unbearable, the real
danger may not be the kidnapper but someone
much closer to home . . .

Coming soon

That Morning

The wheels on the bus went round and round, high in the air, until a weighty crust of mud brought them to an abrupt halt. Above was a flock of leaves. Around was a wake of silence. Below was an earthly reminder of the gravity that had been put to the test. It won. The minibus lay on its side on the forest floor.

The girl too. She rolled her head. The fingers of a yew tree caught her hair. A sluggish raindrop landed on her eyelid. She startled awake. Beyond the massive tree trunk, the bus blurred into focus.

Why is it there? Why am I out here? Why is my arm bent that way? Why—

The bus. The money. The man.

Her mind threw open its shutters to a winter morning. The stark return of everything that had happened already that day, everything that'd changed, everything that was still to be decided. She needed to get up, move, find the others, right now, because they needed her help. She inched one foot through the mulch until her leg was straight. Other foot. The left arm, judging by the firework shooting into her shoulder, was going to be a problem.

There was a creak and crash inside the bus.

Someone else moving.

As quickly as the pain flared, it faded. Perhaps a fracture, then, not a break. It could still send her into shock. Couldn't afford that, needed to think. She rolled onto her knees and up onto her feet, cupping the arm across her chest. *Adrenaline numbs pain, aids survival, fight or flight*, she knew this from her studies. She was going to get the grades to get into medicine to get the life she wanted. *And to get through this day.* But she didn't know how long she had until the pyrotechnics returned.

The minibus was at the bottom of a deeply gouged bank. She staggered around the vehicle, its underneath slick and complex and filthy, as obscenely fascinating as entrails. A few feet beyond, the windscreen lay in one piece. Except it had been shattered by the body crumpled in the centre of the glass. Blood pooled around the figure. It looked unreal . . . it looked dead.

'We need to make sure.' A quiet voice from behind jolted her.

'Can you make sure?' She tapped his arm to encourage him forward. 'Please?'

He walked onto the glass, slip-sliding on foal legs, but managed to keep his balance in the mess. He took stock, levelled himself, then slammed his right boot into the figure's chin. When the body settled, facing the other way now, he stepped back off the windscreen. He wiped his right boot on the fallen leaves. When that was done, his eyes roamed her face.

'You're pale.'

'My arm might be broken.'

Slushy footsteps as another person arrived.

'Where's—?' The voice hitched an octave. 'Oh. My. God.'

Pain radiated from her shoulder up her neck to form a hot halo inside her skull.

It was starting.

The newcomer pointed at the windscreen. 'Are we sure they're—?'

'Oh, yeah,' said the boy. 'We're sure now.'